Selected Titles in This Series

680 **Joachim Zacharias,** Continuous tensor products and Arveson's spectral C^*-algebras, 2000

679 **Y. A. Abramovich and A. K. Kitover,** Inverses of disjointness preserving operators, 2000

678 **Wilhelm Stannat,** The theory of generalized Dirichlet forms and its applications in analysis and stochastics, 1999

677 **Volodymyr V. Lyubashenko,** Squared Hopf algebras, 1999

676 **S. Strelitz,** Asymptotics for solutions of linear differential equations having turning points with applications, 1999

675 **Michael B. Marcus and Jay Rosen,** Renormalized self-intersection local times and Wick power chaos processes, 1999

674 **R. Lawther and D. M. Testerman,** A_1 subgroups of exceptional algebraic groups, 1999

673 **John Lott,** Diffeomorphisms and noncommutative analytic torsion, 1999

672 **Yael Karshon,** Periodic Hamiltonian flows on four dimensional manifolds, 1999

671 **Andrzej Rosłanowski and Saharon Shelah,** Norms on possibilities I: Forcing with trees and creatures, 1999

670 **Steve Jackson,** A computation of δ_5^1, 1999

669 **Seán Keel and James McKernan,** Rational curves on quasi-projective surfaces, 1999

668 **E. N. Dancer and P. Poláčik,** Realization of vector fields and dynamics of spatially homogeneous parabolic equations, 1999

667 **Ethan Akin,** Simplicial dynamical systems, 1999

666 **Mark Hovey and Neil P. Strickland,** Morava K-theories and localisation, 1999

665 **George Lawrence Ashline,** The defect relation of meromorphic maps on parabolic manifolds, 1999

664 **Xia Chen,** Limit theorems for functionals of ergodic Markov chains with general state space, 1999

663 **Ola Bratteli and Palle E. T. Jorgensen,** Iterated function systems and permutation representation of the Cuntz algebra, 1999

662 **B. H. Bowditch,** Treelike structures arising from continua and convergence groups, 1999

661 **J. P. C. Greenlees,** Rational S^1-equivariant stable homotopy theory, 1999

660 **Dale E. Alspach,** Tensor products and independent sums of $\mathcal{L}_p$-spaces, $1 < p < \infty$, 1999

659 **R. D. Nussbaum and S. M. Verduyn Lunel,** Generalizations of the Perron-Frobenius theorem for nonlinear maps, 1999

658 **Hasna Riahi,** Study of the critical points at infinity arising from the failure of the Palais-Smale condition for n-body type problems, 1999

657 **Richard F. Bass and Krzysztof Burdzy,** Cutting Brownian paths, 1999

656 **W. G. Bade, H. G. Dales, and Z. A. Lykova,** Algebraic and strong splittings of extensions of Banach algebras, 1999

655 **Yuval Z. Flicker,** Matching of orbital integrals on $GL(4)$ and $GSp(2)$, 1999

654 **Wancheng Sheng and Tong Zhang,** The Riemann problem for the transportation equations in gas dynamics, 1999

653 **L. C. Evans and W. Gangbo,** Differential equations methods for the Monge-Kantorovich mass transfer problem, 1999

652 **Arne Meurman and Mirko Primc,** Annihilating fields of standard modules of $\mathfrak{sl}(2,\mathbb{C})^\sim$ and combinatorial identities, 1999

651 **Lindsay N. Childs, Cornelius Greither, David J. Moss, Jim Sauerberg, and Karl Zimmermann,** Hopf algebras, polynomial formal groups, and Raynaud orders, 1998

650 **Ian M. Musson and Michel Van den Bergh,** Invariants under Tori of rings of differential operators and related topics, 1998

(Continued in the back of this publication)

Continuous Tensor Products and Arveson's Spectral C*-Algebras

of the
American Mathematical Society

Number 680

Continuous Tensor Products and Arveson's Spectral C*-Algebras

Joachim Zacharias

American Mathematical Society
Providence, Rhode Island

1991 *Mathematics Subject Classification.*
Primary 46Cxx, 46L40, 46L55, 46L57, 46L80.

Library of Congress Cataloging-in-Publication Data

Zacharias, Joachim, 1966–
    Continuous tensor products and Arveson's spectral C^*-algebras / Joachim Zacharias.
      p. cm. — (Memoirs of the American Mathematical Society, ISSN 0065-9266 ; no. 680)
    "January 2000, volume 143, number 680 (second of 4 numbers)."
    Includes bibliographical references.
    ISBN 0-8218-1545-8 (alk. paper)
    1. C^*-algebras. 2. Tensor products I. Title. II. Series.
QA3.A57 no. 680
[QA326]
510 s—dc21
[512′.55]
                                                                                                                                                    99-054531

Memoirs of the American Mathematical Society

    This journal is devoted entirely to research in pure and applied mathematics.

    Subscription information. The 2000 subscription begins with volume 143 and consists of six mailings, each containing one or more numbers. Subscription prices for 2000 are $466 list, $419 institutional member. A late charge of 10% of the subscription price will be imposed on orders received from nonmembers after January 1 of the subscription year. Subscribers outside the United States and India must pay a postage surcharge of $30; subscribers in India must pay a postage surcharge of $43. Expedited delivery to destinations in North America $35; elsewhere $130. Each number may be ordered separately; *please specify number* when ordering an individual number. For prices and titles of recently released numbers, see the New Publications sections of the *Notices of the American Mathematical Society*.

    Back number information. For back issues see the *AMS Catalog of Publications*.

    Subscriptions and orders should be addressed to the American Mathematical Society, P. O. Box 5904, Boston, MA 02206-5904. *All orders must be accompanied by payment*. Other correspondence should be addressed to Box 6248, Providence, RI 02940-6248.

    Copying and reprinting. Individual readers of this publication, and nonprofit libraries acting for them, are permitted to make fair use of the material, such as to copy a chapter for use in teaching or research. Permission is granted to quote brief passages from this publication in reviews, provided the customary acknowledgment of the source is given.

    Republication, systematic copying, or multiple reproduction of any material in this publication is permitted only under license from the American Mathematical Society. Requests for such permission should be addressed to the Assistant to the Publisher, American Mathematical Society, P. O. Box 6248, Providence, Rhode Island 02940-6248. Requests can also be made by e-mail to reprint-permission@ams.org.

    Memoirs of the American Mathematical Society is published bimonthly (each volume consisting usually of more than one number) by the American Mathematical Society at 201 Charles Street, Providence, RI 02904-2294. Periodicals postage paid at Providence, RI. Postmaster: Send address changes to Memoirs, American Mathematical Society, P. O. Box 6248, Providence, RI 02940-6248.

    © 2000 by the American Mathematical Society. All rights reserved.
This publication is indexed in *Science Citation Index*®, *SciSearch*®, *Research Alert*®, *CompuMath Citation Index*®, *Current Contents*®/*Physical, Chemical & Earth Sciences*.
Printed in the United States of America.

    ∞ The paper used in this book is acid-free and falls within the guidelines
established to ensure permanence and durability.
Visit the AMS home page at URL: http://www.ams.org/

        10 9 8 7 6 5 4 3 2 1   05 04 03 02 01 00

Contents

1	**Introduction**		1
2	**Continuous Tensor Products**		9
	2.1	Tensor Decompositions over Boolean Algebras	9
		2.1.1 Definition	9
		2.1.2 Continuous Tensor Decompositions	11
		2.1.3 Discrete Examples	13
		2.1.4 Continuous Examples	17
	2.2	The Generalized Araki-Woods Theorem	26
		2.2.1 An Araki-Woods Theorem for $\mathcal{B}_0(I)$	26
3	**Algebras Associated to Continuous Tensor Products**		**33**
	3.1	Definition of $L^1(T)$ and $\mathcal{A}(T)$	33
		3.1.1 L^1-Sections as Involutive Banach Algebras	34
	3.2	The C^*-Algebra $\mathcal{A}(T)$ for T of Type I	40
		3.2.1 Representations of $\mathcal{A}(T)$	40
		3.2.2 States	43
		3.2.3 Ideals and Exact Sequences	48
	3.3	Automorphisms and Endomorphisms	51
		3.3.1 Ideal Preserving Automorphisms	51
		3.3.2 General Diagonal Morphisms	55
		3.3.3 Generation by Cones	58
		3.3.4 Pedersen Ideal and Infiniteness	61
		3.3.5 The Canonical Automorphic and Endomorphic Actions on $\mathcal{A}_n$	63
	3.4	Homotopy Invariants	64

		3.4.1	K-Theory	64
		3.4.2	The Homotopy Type of the Automorphism Group	65

4 Arveson's Spectral C^*-Algebras 67

    4.1 Product Systems 67

        4.1.1 E_0-Semigroups and Product Systems 67

    4.2 The Spectral C^*-Algebra $C^*(E)$ of a Product System 75

        4.2.1   The Wiener Hopf C^*-Algebra 75

        4.2.2   The Involutive Banach Algebra $L^1(K_E)$ 80

        4.2.3 $C^*(E)$ and its Universal Property 86

        4.2.4   The C^*-Algebras $\mathcal{W}_n$ 91

    4.3 $C^*(E_n)$ as a Crossed Product 95

        4.3.1   The Banach Algebra Crossed Product $L^1(\mathbb{R}, L^1(T_n))$. 95

        4.3.2   Morita Equivalence between $\mathbb{R} \rtimes \mathcal{A}_n$ and $C^*(E_n)$ 98

        4.3.3   Simplicity 104

        4.3.4   Infiniteness 106

Appendix 108

  A. Bochner Integrals 108

  B. Direct Integrals 111

  C. Conditionally Positive Definite Functions 114

References 115

Abstract

We introduce a notion of continuous tensor products of Hilbert spaces which is closely related to Arveson's product systems and Araki and Woods' Boolean algebras of tensor decompositions. There is a classification into type I, II and III, and we can extend a result of Araki and Woods which classifies the type I case as the symmetric Fock space over a direct integral of Hilbert spaces. There are also easy examples of type III. To any continuous tensor product T we associate a nonsimple C^*-algebra $\mathcal{A}(T)$ and show it to be stably projectionless nuclear prime and infinite. It may be viewed as a continuous analogue of a UHF-algebra. In case of constant multiplicity n in the direct integral we obtain a sequence $\mathcal{A}_n$ of C^*-algebras which turn out to be KK-contractible. They carry an action of $\mathbb{R}$ such that $\mathbb{R} \rtimes \mathcal{A}_n$ is a dilation of Arveson's spectral algebra $C^*(E_n)$ which is thus also KK-contractible. We show that $E \subset \mathcal{M}(C^*(E))$ in many cases. The computation of generators shows the existence of nontrivial projections in $C^*(E_n)$ which implies them being simple nuclear infinite and KK-contractible.

[0]Supported by the DFG-Forschergruppe: Topologie und Nichtkommutative Geometrie
1991 AMS-subjectclassification: prim.: 46L35, 46L40, 46L55, 46L57, 46L80; sec.: 46C
Keywords and Phrases : continuous tensor products , continuous UHF-algebras , semigroups , spectral algebras , K-theory

Chapter 1

Introduction

In a series of papers [Ar 89a]-[Ar 91] Arveson defined certain C^*-algebras which he called spectral C^*-algebras. They are continuous analogues of the Cuntz algebras, and he showed that they are simple and nuclear in many cases. Their representation theory is closely related to semigroups of $*$-endomorphisms on $\mathcal{B}(H)$.

It is pointed out in [Ar 91] and [Ar 94a] that besides that not much more is known about them. In particular, their K-theory is completely unknown.

In this work we establish a crossed product representation of the spectral algebras over the group $\mathbb{R}$ in the easiest cases using a new sort of nonsimple C^*-algebras associated to continuous tensor products of Hilbert spaces.

It follows from this crossed product representation that the spectral algebras have trivial K-theory (they are even KK-contractible) in the mentioned cases.

In order to give a more complete description of the content of this work and to put background material into perspective, we summarize parts of the work of Arveson, Powers, Price etc. in A.-C. and the content of this work in D.-F.

A. E_0-semigroups

In [Po 88] Powers developed an index theory for E_0-semigroups of $*$-endomorphisms of a von Neumann algebra M acting on a separable Hilbert

---

Received by the editor September 2, 1997

space H. An E_0-semigroup is a semigroup of unital normal $*$-endomorphisms of M s.t. $\alpha_0 = id$ and $\mathbb{R}_+ \ni t \mapsto \langle \xi, \alpha_t(A)\eta \rangle$ is continuous for arbitrary $\xi, \eta \in H$ and $A \in M$. Besides conjugacy of E_0-semigroups (which is just unitary equivalence if $M = \mathcal{B}(H)$), there is an important weaker relation namely cocycle conjugacy (c.f.4.1.2.(ii)). If there exists a strongly continuous semigroup of isometries $U(t)$ on the Hilbert space H s.t. $\alpha_t(A)U(t) = U(t)A \quad \forall A \in M \quad t > 0$ and $M = \mathcal{B}(H)$, then, using the generators of α and $U(t)$, Powers defines a representation of M (on the defect space of the generator of $U(t)$). The multiplicity of this representation is defined to be the (numerical) index. He also gives an example of an E_0-semigroup having index 1 (the CAR-flow) and from that one obtains, by taking tensor products, examples of arbitrary index. On the other hand Powers [Po 89], [Po 94] has given a very complicated example of an E_0-semigroup without intertwining semigroups of isometries and one having only 'a few' of them (called type III respectively type II). A result of Arveson ([Ar 89b]) implies that, in this terminology, E_0-semigroups behave exactly like Type I, II and III v. Neumann algebras under taking tensor products. The structure of the higher types of such E_0-semigroups is an important open problem. It is conjectured that there are uncountably many not cocycle conjugate type II and type III E_0-semigroups which would imply that the set of isomorphy classes of product systems defined below is uncountable.

B. Product Systems

In [Ar 89] Arveson introduced the notion of a product system in order to create a spectral theory for E_0-semigroups. The prototype of a product system is given by the collection $E_\alpha = \bigcup_{t>0}\{t\} \times E_\alpha(t)$ where $E_\alpha(t) := \{T \in \mathcal{B}(H) | \alpha_t(A)T = TA \quad \forall A \in \mathcal{B}(H)\}$ and α is an E_0-semigroup on $\mathcal{B}(H)$. Each $E_\alpha(t)$ is a Hilbert space in $\mathcal{B}(H)$ and (if α is not automorphic) $E_\alpha(t) \cap E_\alpha(s) = \{0\}$ for $t \neq s$. Thus E_α is a Borel fibration over $(0, \infty)$ which turns out to be trivial. It is a multiplicatively graded set in the following sense: $A \in E_\alpha(t)$ and $B \in E_\alpha(s)$ implies $AB \in E_\alpha(s+t)$ and $A \otimes B \in E_\alpha(t) \otimes E_\alpha(s) \to AB \in E_\alpha(s+t)$ extends to an isomorphism of Hilbert spaces. These properties are the ingredients for the abstract definition of product systems (c.f. sec.4.1). Product systems form a category and it can then be said what representations of product systems are. Each representation ϕ of a product system E generates a semigroup α s.t. $\phi(E) = E_\alpha$. It can also be shown that each abstract product system is given by an E_0-semigroup ([Ar 90c]). For E_0-semigroups α and β the cocycle conjugacy implies isomorphy of E_α and E_β and conversely ([Ar 89]). Thus classifying E_0-semigroups up to cocycle conjugacy is equivalent to classifying product systems up to isomorphy.

The definition of the index is also more natural using product systems: The intertwining semigroup of isometries $U(t)$ corresponds to a normalized multiplicative section in E_α. The set of multiplicative sections $\mathcal{U}_E$ (also called units) can then be equiped with a certain scalar product. The index is the dimension of the corresponding Hilbert space. It is shown in [PoPr 90] that the two definitions of the index coincide. In particular, the numerical index is independent of the choice of the intertwining semigroup.

The only known explicit examples of product systems are the exponential ones E_n (c.f.sec.4.1.1.A). The corresponding E_0-semigroups have numerical index n. The examples of type II and III E_0-semigroups mentioned above show that there are more interesting product systems. Their explicit structure, however, is unknown.

There is also the notion of decomposable product systems (c.f.Def.4.1.9) which can be shown to be exponential [Ar 94b].

C. Spectral Algebras

In [Ar 90a] Arveson introduced the spectral C^*-algebra $C^*(E)$ of a product system E. Its basic property established there is that nondegenerate representations of $C^*(E)$ are in one to one correspondence with representations of E. It follows that nondegenerate representations of $C^*(E)$ correspond bijectively to semigroups α for which E_α is isomorphic to E. This justifies the name spectral C^*-algebra. Because n is an isomorphism invariant for E_n it is conjectured in [Ar 90d] that the $C^*(E_n)$ are pairwise nonisomorphic.

Furthermore, the spectral algebra $C^*(E)$ is simple if E contains a unit and may be thought of as a continuous analogue of the Cuntz algebras $\mathcal{O}_n$

for the following reason:

Let $H = \mathbb{C}^n$. Then $\mathcal{E} := \bigcup_{i \geq 0} H^{\otimes^i}$ is a "fibration" over $\mathbb{N}$ with fibers equal to $H^{\otimes^i}$. If we define a product in $\mathcal{E}$ by $H^{\otimes^i} \times H^{\otimes^j} \ni (x_1, x_2) \mapsto x_1 \otimes x_2 \in H^{\otimes^{i+j}}$, we get a discrete version of a product system. A representation of $\mathcal{E}$ is a map $\varphi : \mathcal{E} \to \mathcal{B}(H)$ s.t. $\varphi(x)\varphi(y) = \varphi(xy)$ for any $x, y \in \mathcal{E}$ and $\varphi(x)^*\varphi(x') = \langle x, x'\rangle \mathbf{1}$ for $x, x' \in H^{\otimes^i}$. In this case φ is determined by its restriction to $H \subseteq \mathcal{E}$. If (e_i) is an onb of H, we obtain an endomorphism $\rho_\varphi(A) := \sum_i \varphi(e_i) A \varphi(e_i)^*$ (i.e. a discrete semigroup). Thus representations of $\mathcal{E}$ and $\mathcal{E}_n$, the extensions of $\mathcal{O}_n$ by the compacts, are in one to one correspondence.

Arveson's definition of $C^*(E)$ may be illustrated by the discrete case: Let $\mathcal{P}_n$ be the polynomial $*$-subalgebra of $\mathcal{O}_n$ algebraically generated by $\{s_i | \, i = 1, \ldots, n\}$, $s_i^* s_j = \delta_{ij}$, $\sum s_i s_i^* = 1$. Each element of $\mathcal{P}_n$ may be written as

$$\sum a_{\mu\nu} s_\mu s_\nu^* = \sum_{i,j} \left(\sum_{|\mu|=i, |\nu|=j} a_{\mu\nu} s_\mu s_\nu^* \right)$$

If $e_{\mu,\nu}$ is the matrix unit to $e_\mu \in H^{\otimes^i}$ and $e_\nu \in H^{\otimes^j}$, we have the finite rank operator

$$K_{ij} = \sum_{|\mu|=i, |\nu|=j} a_{\mu\nu} e_{\mu,\nu} : H^{\otimes^j} \to H^{\otimes^i}$$

For such families of finite rank operators the product in $\mathcal{P}_n$ turns out to be given by the formula

$$(KL)_{ij} = \sum_p \left[\sum_{l=0}^{min(i,p)} (K_{i-l,p-l} \otimes \mathbf{1}_{H^{\otimes l}}) L_{pj} + \sum_{l=1}^{min(p,j)} K_{ip}(L_{p-l,j-l} \otimes \mathbf{1}_{H^{\otimes l}}) \right]$$

Thus the enveloping C^*-algebra of the algebra of compact kernels with multiplication as above is $\mathcal{E}_n$. In [Ar 90a] Arveson defined the spectral C^*-algebra $C^*(E)$ as the enveloping C^*-algebra of the Banach algebra formed by L^1-sections of compact kernels with an analogous product where the sums are replaced by integrals (c.f. sec.4.2.2).

D. Continuous Tensor Products (Ch.2)

Starting with an abstract product system E the Hilbert space $E(t)$ for a given $t > 0$ corresponds to the i-th tensor power $H^{\otimes^i}$ and is thus a sort of continuous tensor product. There are some notions of continuous tensor products in the older literature. The one we use is similar to Araki and Woods' Boolean algebras of tensor decompositions (c.f. [ArWo 66], [Gui 72, Ch.5]). We do not assume completeness of the Boolean algebra, no vacuum vector, no infinite partitions and infinite tensor products.

The idea is that a tensor decomposition over a Boolean algebra $\mathcal{B}$ is a Hilbert space decomposing into the tensor product $\bigotimes_{\Delta \in P} \mathcal{H}_\Delta$ of 'local' Hilbert spaces $\mathcal{H}_\Delta$ for each finite partition of $1 \in \mathcal{B}$. Although we only consider very few examples of Boolean algebras they seem to be a convenient language. As with product systems, tensor decompositions over a fixed Boolean algebra form a category. The two notions are connected by the fact that $E(t)$ is a tensor decomposition over the Boolean interval algebra $\mathcal{B}_0(0,t)$ generated by the halfopen subintervals of $(0,t)$ for any $t > 0$. For E a product system, $E(t)$ is a continuous tensor decomposition for each $t > 0$. If E and F are isomorphic, then $E(t)$ and $F(t)$ are equivalent for all $t > 0$.

The phenomenon which occurs is that there needn't be any 'pure tensors' (c.f. 2.1.1.(iii)). Following Powers terminology for E_0-semigroups we call tensor decompositions type I, II or III according to how many pure tensors they contain. If E is a product system, then $E(t)$ is type I, II, III iff the same holds for E. $E_\alpha(t)$ in Powers' example [Po 89] is therefore of type III as a tensor decomposition.

Using a trivial rearrangement (c.f. 2.1.3.C and 2.1.4.C), we obtain uncountably many inequivalent continuous tensor decompositions over $\mathcal{B}_0(0,t)$. It shows that tensor decompositions over noncomplete Boolean algebras have at least the same complexity as ITPFI-factors. This might give a hint in constructing other non type I E_0-semigroups.

We generalize a theorem of Araki and Woods to the situation of the Boolean interval algebra to show that each type I continuous tensor decomposition is exponential. This provides a short proof of the main result in [Ar 94b].

E. The C^*-Algebras $\mathcal{A}(T)$ (Ch.3)

We associate a non simple C^*-algebra $\mathcal{A}(T)$ to a continuous tensor product T and analyse them for T of type I. An essential tool is the computation of the state space from which we obtain many other objects. We prove directly that they are stably projectionless non-AF infinite nontype I nuclear prime C^*-algebras and there is a decreasing family of ideals. Regarding a certain subset of the pure states allows us to show that if there exists a diagonal isomorphism (i.e. one preserving canonical complements of the ideals) between any two of them, then the associated direct integrals of the tensor decompositions must be (essentially) unitarily equivalent. If the direct integral is just $L^2(I, \mathbb{C}^n)$, the algebras are called $\mathcal{A}_n$. Then they can also be thought of as the diagonal algebra of Arveson's algebra of compact kernels mentioned above. $\mathcal{A}_n$ can be generated by n+1 algebras isomorphic to $C_0(0, 1]$. Note in this circumstance that a UHF-algebra of type $(n+1)^\infty$ is (up to compacts) a reduced free product of n+1 algebras isomorphic to $c = C(\mathbb{N} \cup \infty)$ and $\mathcal{F}^\infty$ is an infinite free product of them. It is easy to show that $\mathcal{A}_n$ is KK-contractible. The K-theory of $\mathcal{A}(T)$ in the general case is not known.

F. The Crossed Product Representation (Ch.4)

The structure of the Banach algebras described above under point C. remains mysterious in [Ar 90a]. Partly for this reason the proof of the spectral properties is quite involved. A better understanding of them could be useful also for the question of pure infiniteness. As a first step we can compute approximate units (4.2.14). In case of $E = E_n$ it is possible to dilate Arveson's Banach algebra $L^1(K_{E_n})$ to a Banach algebra crossed product over $\mathbb{R}$. It turns out that $C^*(E_n)$ is Morita equivalent to $\mathbb{R} \rtimes \mathcal{A}_n$ for a canonical action α of $\mathbb{R}$ on $\mathcal{A}_n$ which means that $C^*(E_n)$ is a corner in $\mathbb{R} \rtimes \mathcal{A}_n$, in complete analogy to $\mathcal{O}_n$. Note that our proof seems to need the approximate unit in $L^1(K_{E_n})$ and such a property is perhaps necessary in order to have a good dilation theory for topological semigroup crossed products. We give a proof of the simplicity of $\mathbb{R} \rtimes \mathcal{A}_n$ which follows Arveson's proof of the simplicity of $C^*(E_n)$.

In [Ar 91] it is shown that integrals over semigroups of isometries with a certain relation generate C^*-algebras $\mathcal{W}_n$ which are Morita equivalent to $C^*(E_n)$. It turns out that $C^*(E_n) \cong \mathcal{W}_n$ and there is a relatively short proof of this fact.

We also show that if E is a product system containing a unit, then E is naturally contained in the twosided multiplier algebra of $C^*(E)$ and not only in the leftmultipliers. So we have many isometries in $\mathcal{M}(C^*(E))$. For the proof we use a property of the Wiener-Hopf-C^*-algebra already observed in [Ar 90a] (without proof).

It is easy to compute nontrivial projections in $C^*(E_n)$ which implies together with $K_0(C^*(E_n)) = 0$ that there exists an infinite projection in a matrix algebra over $C^*(E_n)$. This is one possible weak form of infiniteness. Thus $C^*(E_n)$ are separable simple nuclear infinite and KK-contractible C^*-algebras. If we could prove them being purely infinite, then, by Kirchberg's classification [Ki 94] of purely infinite simple C^*-algebras (see also [Ph 95]), they would all be isomorphic to $\mathcal{O}_2 \otimes \mathcal{K}$. If any two of them would turn out to be not isomorphic to each other, they would provide the first example of an infinite but not purely infinite nuclear and simple C^*-algebra.

Terminological Remark

We assume all Hilbert spaces to be separable, the scalar product being conjugate linear in the first variable and mean by onb an orthonormal basis. $\Theta_{\xi,\eta}$ denotes the rank one operator $\zeta \mapsto \langle \eta, \zeta \rangle \xi$. The onesided shift on $L^2(\mathbb{R}_+)$ is denoted by T_t and the twosided shift on $L^2(\mathbb{R})$ by S_t. Often also their higher multiplicity counterparts are denoted by the same letters. For a subspace of a tensor product $A \otimes B$ of the form $\mathbb{C}a \otimes B'$, $B' \subseteq B$ a subspace we write $a \otimes B'$. The closed linear span of a set S is denoted by $[S]$. $A^\sim$ is the unitization etc.

Acknowledgement

The present article is (essentially) the author's Ph.D. thesis (Heidelberg 1996) and he wishes to thank Professor J.Cuntz for many helpful remarks and patience. Without his support the present work wouldn't exist. Thanks are also due to priv.Doz.Dr.R.Speicher for taking the second opinion and to E.Blanchard for helpful discussions.

This work is essentially a continuation of Arveson's whose papers have been highly beneficial for us. We have been borrowing many of his ideas and technical ground work.

Furthermore, we want to thank B.Neubüser for a lot of help and all individuals who supported this project.

Chapter 2

Continuous Tensor Products

2.1 Tensor Decompositions over Boolean Algebras

2.1.1 Definition

Recall ([Gui 72], [Hal 67]) that a Boolean algebra is a distributive complemented lattice with smallest and biggest elements 0 and 1. Each Boolean algebra is the algebra of clopen sets of a zero dimensional compact space (Stone's representation theorem) and hence can be thought of as an algebra of sets (with some care). We only consider:

- b_0 the Boolean algebra given by the finite subsets of $\mathbb{N}$ and its complements.

- b_1 the complete Boolean algebra consisting of all subsets of $\mathbb{N}$.

- $\mathcal{B}_0(I)$ the Boolean algebra consisting of the finite unions of halfopen intervals $[x,y) \cap I$, $I = (a,b)$, $x, y, a, b \in \mathbb{R} \cup \{-\infty, \infty\}$.

- $\mathcal{B}(I)$ the Borel subsets of I.

We now define tensor decompositions in (i), establish them as a category in (ii) and introduce, following Powers' terminology for E_0-semigroups (c.f.[Po 94]), a type classification in (iii). A priori this type classification has nothing to do with types of factors.

Definition 2.1.1 *(i) A tensor decomposition $T = (\mathcal{H}, \mathcal{B}, (\mathcal{H}_\Delta)_{\Delta \in \mathcal{B}}, \varphi_P, \varphi_{PQ})$ (over $\mathcal{B}$) is given by*

  (a) *A Hilbert space $\mathcal{H}$*

  (b) *A Boolean algebra $\mathcal{B}$ with smallest and biggest elements 0 and 1*

  (c) *A family of Hilbert spaces $(\mathcal{H}_\Delta)_{\Delta \in \mathcal{B}}$ s.t. $\mathcal{H}_0 = \mathbb{C}$, $\mathcal{H}_1 = \mathcal{H}$*

  (d) *Unitaries φ_P, φ_{PQ} indexed by finite partitions s.t. for $P \leq Q$* [1]

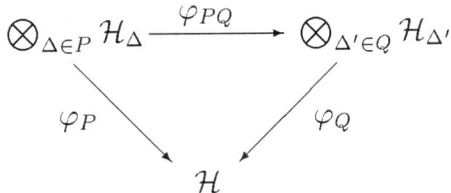

  commutes and all maps preserve tensors (i.e. $\varphi_{PQ}(\otimes \xi_\Delta) = \otimes \xi'_{\Delta'}$)

(ii) A morphism between tensor decompositons $T_i = (\mathcal{H}^i, \mathcal{B}, \varphi^i)$ over $\mathcal{B}$ is a bounded linear map $\Phi : \mathcal{H}^1 \to \mathcal{H}^2$ s.t. for each finite partition P there are bounded linear maps $\Phi_\Delta : \mathcal{H}^1_\Delta \to \mathcal{H}^2_\Delta$ with the property $\Phi(\varphi^1_P(\otimes \xi_\Delta)) = \varphi^2_P(\otimes \Phi_\Delta(\xi_\Delta))$ for all $\xi_\Delta \in \mathcal{H}^1_\Delta$, $\Delta \in P$. Φ is also called factorizable.

(iii) For a tensor decomposition T a vector $\xi \in \mathcal{H}$ is called factorizable if for each finite partition P and $\Delta \in P$ there are $\xi_\Delta \in \mathcal{H}_\Delta$ s.t. $\xi = \varphi_P(\otimes_{\Delta \in P} \xi_\Delta)$. If $\mathcal{F}$ is the set of all of them, then T is called type I, II or III according to whether $[\mathcal{F}] = \mathcal{H}$, $0 \neq [\mathcal{F}] \neq \mathcal{H}$ or $\mathcal{F} = 0$.

Thus a tensor decomposition is a Hilbert space which decomposes in a compatible way into the tensor product of "local" Hilbert spaces. If we replace the tensor product by the direct sum, we obtain essentially the notion of direct integrals.

Remark 2.1.2 *(i) We often write $T = (\mathcal{H}, \mathcal{B})$ or just $\mathcal{H}$ for a tensor decomposition.*

(ii) We usually drop φ_P from notation and write $\otimes \xi_\Delta$ for elements in $\mathcal{H}$.

---
[1] $P \leq Q :\Leftrightarrow \forall \Delta \in P \exists \Delta' \in Q : \Delta \leq \Delta'$

(iii) Isomorphic tensor decompositions are also called equivalent. Any isomorphism may be chosen to be unitary.

(iv) $\hat{T} = (\mathcal{H}, \hat{\mathcal{B}}, \hat{\varphi})$ is called an extension of $T = (\mathcal{H}, \mathcal{B}, \varphi)$ if $\mathcal{B} \subseteq \hat{\mathcal{B}}$ and $\varphi_P = \hat{\varphi}_P$, $\varphi_{PQ} = \hat{\varphi}_{PQ}$ for P, Q partitions in $\mathcal{B}$.

(v) Let $T_k = (\mathcal{H}^k, \mathcal{B}_k, \varphi^k)$, $k = 1, 2$ be tensor decompositions.

If $\mathcal{B}_1 = \mathcal{B}_2 = \mathcal{B}$, then $T_1 \otimes T_2 = (\mathcal{H}^1 \otimes \mathcal{H}^2, \mathcal{B}, (\mathcal{H}^1_\Delta \otimes \mathcal{H}^2_\Delta), \varphi^1_P \otimes \varphi^2_P, \varphi^1_{PQ} \otimes \varphi^2_{PQ})$ is a tensor decomposition.

There is always the tensor decomposition $T_1 \hat{\otimes} T_2 = (\mathcal{H}^1 \otimes \mathcal{H}^2, \mathcal{B}_1 \times \mathcal{B}_2, (\mathcal{H}^1_{\Delta_1} \otimes \mathcal{H}^2_{\Delta_2})_{\Delta_1 \times \Delta_2}, \varphi^1_{P_1} \otimes \varphi^2_{P_2}, \varphi^1_{P_1 Q_1} \otimes \varphi^2_{P_2 Q_2})$ over the product algebra $\mathcal{B}_1 \times \mathcal{B}_2$.

2.1.2 Continuous Tensor Decompositions

One could define continuous tensor decompositions just as being defined over $\mathcal{B}_0(I)$, $I = (a, b)$. However, then also a (discrete) finite or infinite tensor product would be continuous with finitely, resp. countably many points as hidden atoms.

Definition 2.1.3 $T = (\mathcal{H}, \mathcal{B}_0(I))$ *is called weakly continuous if there are onb's $(e_n(t)) \subseteq \mathcal{H}_t := \mathcal{H}_{I \cap [-\infty, t)}$, $(f_m(t)) \subseteq \mathcal{H}^t := \mathcal{H}_{I \cap [t, \infty]}$ s.t. $I \ni t \mapsto e_n(t) \otimes f_m(t)$ are continuous for $n, m \in \mathbb{N}$ as maps from I into $\mathcal{H}$. We also assume $\mathcal{H}_\Delta$ to be infinite dimensional for each nonzero Δ in $\mathcal{B}_0(I)$.*

We can fix an onb (v_m) of $\mathcal{H}$ and define isometries $V_n(t) : v_m \mapsto e_n(t) \otimes f_m(t)$. $\rho_t(A) := \sum_{n=0}^\infty V_n(t) A V_n(t)^*$ is a family of unital endomorphisms associated to the tensor decomposition which is not a semigroup but only monotone in the sense that the image decreases. Another choice of onbs as above leads to a family of endomorphisms of the form $\rho_t \circ Ad(W_t)$ where $I \ni t \mapsto W_t \in \mathcal{B}(\mathcal{H})$ is any strongly continuous family of unitaries. Thus ρ_t is unique up to such a perturbation.

Lemma 2.1.4 *For $T = (\mathcal{H}, \mathcal{B}_0(I))$ a tensor decomposition the following conditions are equivalent:*

(i) T is weakly continuous.

(ii) Any associated family of endomorphisms is strongly continuous.

<u>Proof:</u> (i) $\Rightarrow$ (ii): Define a family of unitaries $I \times I \ni (t_1, t_2) \mapsto \mathcal{B}(\mathcal{H})$ by sending $e_n(t_1) \otimes f_m(t_1)$ to $e_n(t_2) \otimes f_m(t_2)$. Obviously, this map is weakly continuous, and by computing $\|(U(t_1 + \varepsilon, t_2 + \delta) - U(t_1, t_2))\xi\|^2$, we obtain strong continuity. But we also have $\rho_t(A) = U(t, t_0)\rho_{t_0}(A)U(t_0, t)$ for each $A \in \mathcal{B}(\mathcal{H})$.

(ii) $\Rightarrow$ (i): We only need to find a strongly continuous family of unitaries $U(t_1, t_2)$ as in (i) $\Rightarrow$ (ii). To this end follow [Ar 89 Lemma 2.3]. The proof there uses nowhere the semigroup property. The idea is that if $P = \Theta_{v_m, v_m}$ is the minimal projection corresponding to v_m, then $\rho_t(P)\mathcal{H} = \mathcal{H}_t \otimes f_m(t)$ is a continuous family of infinite dimensional subspaces of $\mathcal{H}$ and thus by [Dix 10.8.7] trivial ($(e_n(t))$ is a choice of a trivialization). $\square$

Lemma 2.1.5 If $T = (\mathcal{H}, \mathcal{B}_0(I))$ is weakly continuous, then the family $M_t = \mathbf{1} \otimes \mathcal{B}(\mathcal{H}^t) \subseteq \mathcal{B}(\mathcal{H}) =: M$ is a continuous family of type I subfactors of M in the sense that

$$\overline{\bigcup_{t>t_0} M_t}^w = M_{t_0} = \bigcap_{t<t_0} M_t$$

Moreover, for $t_0 \in I$

$$\bigcap_{\varepsilon>0} M_{[t_0-\varepsilon, t_0+\varepsilon]} = \mathbb{C}\mathbf{1}$$

where $M_{[t_0-\varepsilon, t_0+\varepsilon]} := \mathcal{B}(\mathcal{H}_{[t_0-\varepsilon, t_0+\varepsilon]}) \otimes \mathbf{1}$.

<u>Proof:</u> The first assertion follows from the strong continuity of ρ_t. For the second remark that

$$M_{[t_0-\varepsilon, t_0+\varepsilon]} = M_{t_0-\varepsilon} \cap M'_{t_0+\varepsilon}$$

and (M_t) is continuous (in the above sense) iff (M'_t) is continuous. $\square$

Definition 2.1.6 $T = (\mathcal{H}, \mathcal{B}_0(I))$, $I = (a, b)$ is called continuous if T is weakly continuous and $\bigcap_{t>a} M_t = \mathbb{C}\mathbf{1}$, $\overline{\bigcup_{t<b} M_t}^w = \mathcal{B}(\mathcal{H})$.

Remark 2.1.7 *(i) Let T be continuous. Then $\overline{\bigcup_{t>t_0} M_t}^w = M_{t_0} = \bigcap_{t<t_0} M_t$ and $\bigcap_{t>a} M_t = \mathbb{C}\mathbf{1}$, $\overline{\bigcup_{t<b} M_t}^w = \mathcal{B}(\mathcal{H})$.*

(ii) For $T = (\mathcal{H}, \mathcal{B}_0(I))$ weakly continuous we cannot even expect $\bigcap_{t>a} M_t$ and $\overline{\bigcup_{t<b} M_t}^w$ to be type I factors and it is easy to find examples where they aren't. Thus there are nontype I weakly continuous tensor decompositions.

2.1.3 Discrete Examples

A. Infinite Tensor Products

Let (H_n, Ω_n) be a sequence of Hilbert spaces and $\Omega_n \in H_n$, $\|\Omega_n\| = 1$. The infinite tensor product $\mathcal{H} := \bigotimes_{n \in \mathbb{N}}(H_n, \Omega_n)$ (c.f. [ArWo 66], [Gui 72]) is defined as the inductive limit of the Hilbert spaces $H_F := \bigotimes_{n \in F} H_n$, where $F \subseteq \mathbb{N}$ is a finite subset under the embeddings $H_F \hookrightarrow H_{F'}$, $\xi \mapsto \xi \otimes \Omega_{n_1} \otimes \Omega_{n_2} \otimes \ldots \otimes \Omega_{n_k}$ if $F \subseteq F'$ and $\{n_1, \ldots, n_k\} = F' \setminus F$.

The factorizable vectors are obtained as follows: Recall that for a sequence $(a_n) \subseteq \mathbb{C}$ the infinite product $\prod a_n$ is the limit of the net $\prod_{n \in F} a_n$, $F \subseteq \mathbb{N}$ finite, directed by inclusion. In this case $\prod a_n$ is called convergent. Note that $\prod a_n$ is convergent to $a \neq 0$ iff $\sum_{n=0}^{\infty} \log a_n$ converges absolutely iff $\sum_{n=0}^{\infty} |a_n - 1| < \infty$. A sequence (ξ_n), $\xi_n \in H_n$ s.t. there is $k_0 \in \mathbb{N}$ with the property that the product $\prod_{n>k_0}\langle \xi_n, \Omega_n \rangle$ converges and is nonzero defines a linear functional on the infinite tensor product as follows: $f_F := f|H_F$, $f_F(\zeta_{n_1} \otimes \ldots \otimes \zeta_{n_r}) := \langle \xi_{n_1}, \zeta_{n_1} \rangle \cdots \langle \xi_{n_r}, \zeta_{n_r} \rangle \prod_{\mathbb{N}\setminus F}\langle \xi_m, \Omega_m \rangle$ has the property that $f_{F'}$ extends f_F for $F \subseteq F'$ and $\|f_F\| = \prod_{n \in F} \|\xi_n\| \prod_{m \in \mathbb{N}\setminus F} |\langle \xi_m, \Omega_m \rangle|$. Thus (ξ_n) defines a nonzero vector $\otimes \xi_n$ in $\mathcal{H}$ if $\prod \|\xi_n\|$ and $\prod_{n>k_0}\langle \xi_n, \Omega_n \rangle$ exist and are nonzero for some $k_0 \in \mathbb{N}$. This happens iff $(\otimes_F \xi_n) \otimes (\otimes_{F^c} \Omega_n) \to \otimes \xi_n$ or if

$$\sum |\langle \xi_n, \Omega_n \rangle - 1| < \infty \quad \text{and} \quad \sum |\|\xi_n\| - 1| < \infty$$

Such vectors, called product vectors span $\mathcal{H}$. For any two product vectors $\otimes \xi_n^1$ and $\otimes \xi_n^2$ the product $\prod_n \langle \xi_n^1, \xi_n^2 \rangle$ converges and is equal to $\langle \otimes \xi_n^1, \otimes \xi_n^2 \rangle$. If $\Delta \subseteq \mathbb{N}$ is any subset, we may put $\mathcal{H}_\Delta = \bigotimes_{n \in \Delta}(H_n, \Omega_n)$ and obtain a tensor decomposition over b_1. The following Lemma may be found in [Gui 72 A.3] and [ArWo 66].

Lemma 2.1.8 *The product vectors are exactly the factorizable vectors in $\mathcal{H}$ (as tensor decomposition over b_0 or b_1).*

Proof: Let ξ be a factorizable unit vector. Then taking the sequence of partitions $\{\{0\},\ldots,\{k\},\{k+1,\ldots\}\}$ we get a sequence $\xi_n \in H_n$, $\|\xi_n\|=1$ and $\xi^N \in \mathcal{H}_{\{N+1,\ldots\}}$ s.t. $\xi = \xi_0 \otimes \ldots \otimes \xi_N \otimes \xi^N$. If $K_N := \mathcal{H}_{\{0,\ldots,N\}} = H_0 \otimes \ldots \otimes H_N$ is considered as a subspace of $\mathcal{H}$, then $dist(K_N,\xi) = \|\xi - P_{K_N}\xi\| = \|\xi_0 \otimes \ldots \otimes \xi_N \otimes \xi^N - \xi_0 \otimes \ldots \otimes \xi_N \otimes \langle \Omega^N, \xi^N \rangle \Omega^N\| = \|\xi^N - \langle \Omega^N, \xi^N \rangle \Omega^N\| = dist(\mathbb{C}\Omega^N, \xi^N) \to 0$ for $N \to \infty$ in particular $|\langle \Omega^N, \xi^N \rangle| \to 1$. Let $N_0 \in \mathbb{N}$ be s.t. $|1 - |\langle \Omega^N, \xi^N \rangle|| < 1/2$ for $N > N_0$. We may assume $|\langle \Omega_n, \xi_n \rangle| = \langle \Omega_n, \xi_n \rangle$ and $|\langle \Omega^N, \xi^N \rangle| = \langle \Omega^N, \xi^N \rangle$ for $n, N > N_0$. If $F_i \subseteq \{N_0+1, N_0+2, \ldots\}$, $i=1,2$ are finite subsets, then $|\langle \Omega_{F_1^c}, \xi_{F_1^c} \rangle - 1| \geq |\langle \Omega_{F_2^c}, \xi_{F_2^c} \rangle - 1|$ for $F_1 \subseteq F_2$. Thus for any sequence $F_k \nearrow \{N_0+1, N_0+2, \ldots\}$ the products $\prod_{n \in F_k} \langle \Omega_n, \xi_n \rangle$ converge to $\langle \Omega^{N_0}, \xi^{N_0} \rangle$. The converse is obvious. □

B. The Powers Factors

Recall that the Powers factor $\mathcal{R}_\lambda$, $\lambda \in (0,1)$ is equal to the weak closure of $\pi_{\omega_\lambda}(M(2^\infty))$ where ω_λ is the following product state $\otimes \omega_i^\lambda$ on the UHF-algebra $M(2^\infty)$

$$\omega_i^\lambda = \begin{pmatrix} (1+\lambda)^{-1} & \\ & \lambda(1+\lambda)^{-1} \end{pmatrix}$$

The following is well known:

Lemma 2.1.9 *(i) The GNS-representation of $\otimes M_i = M(2^\infty)$, $M_i = M_2(\mathbb{C})$ is the infinite tensor product space $\bigotimes_{i=0}^\infty (H_{\omega_i^\lambda}, \Omega_{\omega_i^\lambda})$, $(H_{\omega_i^\lambda}, \Omega_{\omega_i^\lambda})$ the GNS-space with cyclic vector $\Omega_{\omega_i^\lambda}$ of ω_i^λ on M_i, $\pi_{\omega_\lambda} = \otimes \pi_{\omega_i^\lambda}$ i.e. $\pi_{\omega_\lambda}(a_1 \otimes a_2 \otimes \ldots a_k) = \pi_{\omega_1^\lambda}(a_1) \otimes \ldots \otimes \pi_{\omega_k^\lambda}(a_k) \otimes \mathbf{1}$, $\Omega_{\omega_\lambda} = \otimes \Omega_{\omega_i^\lambda}$.*

(ii) If $M_i = \mathcal{B}(H_i)$, H_i separable and infinite dimensional, let $\mathcal{A}$ be the $*$-algebra which is the inductive limit of $M_1 \otimes \ldots \otimes M_k$, $k \to \infty$, $\omega_i \in M_{i,*}$ s.t. ω_i has the eigenvalues $(1+\lambda)^{-1}$ and $\lambda(1+\lambda)^{-1}$. Let ω be the functional on $\mathcal{A}$ defined as $\omega = \lim_{k\to\infty} \omega_1 \otimes \ldots \otimes \omega_k$. Then the corresponding GNS-space is isomorphic to the infinite tensor product $\bigotimes (H_{\omega_i}, \Omega_{\omega_i})$, $(H_{\omega_i}, \Omega_{\omega_i})$ the GNS-space of ω_i on M_i. $\pi_\omega(\mathcal{A})''$ is unitarily equivalent to $\mathcal{R}_\lambda \otimes \mathcal{B}(H)$.

Proof: (i): For $a_1 \in M_1, \ldots, a_k \in M_k$ the map

$$H \ni \pi_{\omega_\lambda}(a_1 \otimes \ldots \otimes a_k)\Omega_{\omega_\lambda} \mapsto \pi_{\omega_1^\lambda}(a_1)\Omega_{\omega_1^\lambda} \otimes \ldots \otimes \pi_{\omega_k^\lambda}(a_k)\Omega_{\omega_k^\lambda}$$

is scalar product preserving, sends Ω_{ω_λ} to $\Omega_{\omega_1^\lambda} \otimes \ldots \otimes \Omega_{\omega_k^\lambda}$ and thus extends to a unitary $H_{\omega_\lambda} \to \otimes(H_{\omega_i^\lambda}, \Omega_{\omega_i^\lambda})$ by the universal property of inductive limits.

(ii): The same argument: Write $M_i = M_{i,1} \otimes M_{i,2}$, $M_{i,1} = M_2(\mathbb{C})$ and $M_{i,2} = \mathcal{B}(H_{i,2})$. There is a vector state ρ_i on $M_{i,2}$ s.t. $\omega_i = \omega_i^\lambda \otimes \rho_i$, ω_i^λ as in (i). We have a unitary equivalence $\pi_{\omega_i} \cong \pi_{\omega_i^\lambda} \otimes \pi_{\rho_i}$ and the map

$$\pi_\omega((a_{11} \otimes a_{12}) \otimes (a_{21} \otimes a_{22}) \otimes \ldots \otimes (a_{k1} \otimes a_{k2}))\Omega_\omega \mapsto$$

$$\pi_{\omega_1}(a_{11} \otimes a_{12})\Omega_{\omega_1} \otimes \ldots \otimes \pi_{\omega_k}(a_{k1} \otimes a_{k2})\Omega_{\omega_k}$$

$a_{i1} \in M_{i,1}$, $a_{i,2} \in M_{i,2}$, $i = 1, \ldots, k$ extends to the desired unitary equivalence. $\square$

Remark 2.1.10 *(i) $\pi_{\omega_\lambda}(M(2^\infty))''$ is the hyperfinite II_1-factor for $\lambda = 1$. We include this case as $\mathcal{R}_1$.*

(ii) Powers [Po 67] has shown that the factors $\mathcal{R}_\lambda$ are pairwise nonisomorphic of type III for $\lambda \in (0,1)$.

C. Rearranged Infinite Tensor Products

Let $\mathcal{H}^\lambda = \bigotimes_{n \in \mathbb{N}}(H_n, \Omega_n^\lambda)$ be the infinite tensor product space with $H_n = \mathbb{C}^4$. We write $\mathbb{C}^4$ as $\mathbb{C}^2 \otimes \mathbb{C}^2 = H_{(n,0)} \otimes H_{(n,1)}$ in some way. Using a bijection between $\{(n,i)|n \in \mathbb{N}, i \in \{0,1\}\}$ and $\mathbb{N}$, we may rewrite $\mathbb{C}^4 \otimes \mathbb{C}^4 \otimes \ldots = \bigotimes(H_n, \Omega_n^\lambda) = \mathbb{C}^2 \otimes \mathbb{C}^2 \otimes \ldots$ as a rearranged infinite tensor product.

Specifically, we fix onb's $\{e_1^{(n,i)}, e_2^{(n,i)}\}$ of $H_{(n,i)} = \mathbb{C}^2$ and put $\Omega_n^\lambda = (1+\lambda)^{-\frac{1}{2}} e_1^{(n,0)} \otimes e_1^{(n,1)} + \lambda^{\frac{1}{2}}(1+\lambda)^{-\frac{1}{2}} e_2^{(n,0)} \otimes e_2^{(n,1)}$, $\lambda \in (0,1]$. Let $\sigma : \mathbb{N} \to \mathbb{N} \times \{0,1\}$, $\sigma(2k) = (k,0)$, $\sigma(2k+1) = (k,1)$. For $F \subseteq \mathbb{N}$ finite, put $\mathcal{H}^\lambda{}_F = \bigotimes_{n \in F} H_{\sigma(n)}$ and for F cofinite let $r \in \mathbb{N}$ be s.t. $\{2r, 2r+1, \ldots\} \subseteq F$. Then put $\mathcal{H}^\lambda{}_F = \bigotimes_{n \geq r}(H_n, \Omega_n^\lambda) \otimes (\bigotimes_{l \in F \cap \{0, \ldots, 2r-1\}} H_{\sigma(l)})$. $\mathcal{H}^\lambda{}_F$ may be viewed as a subspace of $\mathcal{H}^\lambda$ and is independent of the choice of r.

Let $P = \{\Delta_1, \ldots, \Delta_k\}$ be a partition of $\mathbb{N} = 1 \in b_0$. We take a refinement P' of P of the form $P' = \{\{0\}, \ldots, \{2r-1\}, \{2r, 2r+1, \ldots\}\}$.

For $\xi_0, \ldots, \xi_{2r-1} \in \mathbb{C}^2$ and $\xi_{2r} \in \otimes_{n \geq r}(H_n, \Omega_n^\lambda)$, we have $\otimes_{\Delta \in P}(\otimes_{j \in \Delta} \xi_j) \in \otimes_{\Delta \in P} \mathcal{H}_\Delta$ and such vectors form a total set. We define $\varphi_P(\otimes_{\Delta \in P}(\otimes_{j \in \Delta} \xi_j)) = \xi_0 \otimes \ldots \otimes \xi_{2r}$ and $\varphi_{PQ}(\otimes_{\Delta \in P}(\otimes_{j \in \Delta} \xi_j)) = \otimes_{\Delta' \in Q}(\otimes_{k \in \Delta'} \xi_k)$. Then $(\mathcal{H}^\lambda, b_0)$ is a tensor decomposition.

Proposition 2.1.11 $(\mathcal{H}^\lambda, b_0)$ *are of type III and all mutually inequivalent.*

<u>Proof:</u> Let $\xi \in \mathcal{F}$ be a factorizable unit vector. As we have seen in 2.1.8., the factorizable vectors w.r.t. b_0 in an infinite tensor product are exactly the product vectors. So there are unit vectors $\xi_n \in H_n = \mathbb{C}^4$ s.t. $\xi = \otimes \xi_n$. Each $\xi_n \in H_n = H_{(n,0)} \otimes H_{(n,1)}$ must be a product vector as well, i.e. $\xi_n = \xi_{(n,0)} \otimes \xi_{(n,1)}$. Now for $a, b \in \mathbb{C}^2$, $\Omega^\lambda := \Omega_n^\lambda$ we have

$$dist(\mathbb{C}(a \otimes b), \Omega^\lambda)^2 = 1 - \frac{1}{\|a\|^2 \|b\|^2} |\langle a \otimes b, \Omega^\lambda \rangle|^2$$

$$= 1 - \frac{|\langle \bar{a}, \begin{pmatrix} (1+\lambda)^{-\frac{1}{2}} & \\ & \lambda^{\frac{1}{2}}(1+\lambda)^{-\frac{1}{2}} \end{pmatrix} b \rangle|^2}{\|a\|^2 \|b\|^2}$$

The minimal value of that is

$$1 - max((1+\lambda)^{-1}, \lambda(1+\lambda)^{-1}) = 1 - (1+\lambda)^{-1}$$

Therefore $\sum \|\xi_n - \Omega_n^\lambda\|$ is divergent and because $\|\xi_n\| = 1$ also $\sum |\langle \xi_n, \Omega_n^\lambda \rangle - 1|$ which is a contradiction. So $T_\lambda = (\mathcal{H}^\lambda, b_0)$ is type III. Let M be the von Neumann algebra given by the weak closure of $\mathcal{A} := M_2 \otimes 1 \otimes M_2 \otimes \ldots$. Then for $a \in \mathcal{A}$ and $\Omega^\lambda = \otimes \Omega_n^\lambda$ we have $\langle \Omega^\lambda, a\Omega^\lambda \rangle = (\otimes \omega_i^\lambda)(a)$, ω_i^λ given by

$$\begin{pmatrix} (1+\lambda)^{-1} & \\ & \lambda(1+\lambda)^{-1} \end{pmatrix}$$

as above. Hence M is equal to $\mathcal{R}_\lambda$ and an equivalence of T_λ and $T_{\lambda'}$ would imply a unitary equivalence between $\mathcal{R}_\lambda$ and $\mathcal{R}_{\lambda'}$. □

We remark that there seems to be no way to extend $\mathcal{H}^\lambda$ to a tensor decomposition over b_1. The reason is of course that non type I factors are not spatial. We observe that tensor decompositions over b_0 have at least the same complexity as ITPFI-factors. Tensor decompositions over b_1 are only infinite tensor products (this follows from a result of [ArWo 66]).

CONTINUOUS TENSOR PRODUCTS

2.1.4 Continuous Examples

A. Fock spaces

We review some well known facts about Fock spaces.

Let K be a Hilbert space and $K^{\otimes n} = K \otimes \ldots \otimes K$ be the n-fold tensor product of K with itself. On $K^{\otimes n}$ the permutation group S_n acts by

$$U_\sigma(\xi_1 \otimes \ldots \otimes \xi_n) = \xi_{\sigma 1} \otimes \ldots \otimes \xi_{\sigma n}$$

The symmetric tensor power $K^{\otimes n}_s$ is the subspace of $K^{\otimes n}$ defined as

$$K^{\otimes n}_s := \{\xi \in K^{\otimes n} | U_\sigma \xi = \xi \ \forall \sigma \in S_n\}$$

and is the image of the projection $E = \frac{1}{n!} \sum_{\sigma \in S_n} U_\sigma$. $K^{\otimes n}_s$ is spanned by $\{\xi^{\otimes n} | \xi \in K\}$ because $E(\xi_1 \otimes \ldots \otimes \xi_n) = \frac{1}{n!} \frac{d}{d\lambda_1} \ldots \frac{d}{d\lambda_n}|_{\lambda_1 = \ldots = \lambda_n = 0}(\lambda_1 \xi_1 + \ldots + \lambda_n \xi_n)^{\otimes n}$.

The full Fock space (over K) is the Hilbert space $\mathcal{F}(K) = \bigoplus_{n=0}^\infty K^{\otimes n}$, $K^{\otimes 0} := \mathbb{C}\Omega$, Ω the vacuum and the symmetric Fock space is the subspace $\mathcal{F}^s(K) = \bigoplus_{n=0}^\infty K^{\otimes n}_s$, $K^{\otimes 0}_s := \mathbb{C}\Omega$.

If $\xi \in K$, we define the exponential vector (or exp-vector for short)

$$exp(\xi) := \sum_{n=0}^\infty \frac{1}{\sqrt{n!}} \xi^{\otimes n} \in \mathcal{F}^s(K)$$

($\xi^{\otimes 0} := \Omega$). For $\xi, \eta \in K$ we have $\langle exp(\xi), exp(\eta) \rangle = e^{\langle \xi, \eta \rangle}$.

Lemma 2.1.12 *(i) The map $K \ni \xi \mapsto exp(\xi) \in \mathcal{F}^s(K)$ is continuous and injective.*

(ii) The set of exponential vectors is linearly independent and total.

<u>Proof:</u> (i): $\|exp(\xi) - exp(\eta)\|^2 = e^{\|\xi\|^2} + e^{\|\eta\|^2} - 2Re(e^{\langle \xi, \eta \rangle})$ and injectivity follows from (ii).

(ii): $\frac{1}{\sqrt{n!}} \frac{d^n}{d\lambda^n}|_{\lambda=0} exp(\lambda \xi) = \xi^{\otimes n}$ and $exp(0) = \Omega$ implies $[exp(K)] = \mathcal{F}^s(K)$. For linear independence, following [Par 92], take $\xi_1, \ldots, \xi_k \in K$ all different and suppose there exist $\alpha_1, \ldots, \alpha_k \in \mathbb{C}$ s.t. $\sum_i \alpha_i exp(\xi_i) = 0$. We can find $\xi \in K$ s.t. $\lambda_i := \langle \xi, \xi_i \rangle$ are pairwise different and obtain

$$0 = \left\langle \sum \alpha_i exp(\xi_i), exp(z\xi) \right\rangle = \sum \bar{\alpha}_i e^{z\langle \xi_i, \xi \rangle}$$

identically in $z \in \mathbb{C}$. Substituting $e^z = w$, this means $\sum \bar{\alpha}_i w^{\lambda_i} = 0$ identically in $w \in \mathbb{C} \setminus \{0\}$. □

Corollary 2.1.13 *For each $A \in \mathcal{B}(K)$ there is a unique densely defined linear operator $exp(A)$ on $\mathcal{D} := [exp(\xi) | \xi \in K] \subseteq \mathcal{F}^s(K)$ s.t.*

$$exp(A)exp(\xi) := exp(A\xi)$$

$exp(A)$ *is bounded iff* $\|A\| \leq 1$ *and* $exp(A)\xi^{\otimes n} = (A\xi)^{\otimes n}$. $exp(A)$ *is unitary, isometric or a projection iff A has the corresponding property.*

Proof: By the foregoing Lemma, $exp(A)$ is well defined. By the definition of exp-vectors, $exp(A)\xi^{\otimes n} = (A\xi)^{\otimes n}$. So $\|exp(A)|_{K^{\otimes n}_s}\| = \|A\|^n$ which implies $exp(A)$ to be bounded iff $\|A\| \leq 1$.

Now if $\|A\| \leq 1$, then $\langle exp(A)^* exp(\xi), exp(\eta) \rangle = \langle exp(\xi), exp(A\eta) \rangle = e^{\langle \xi, A\eta \rangle} = e^{\langle A^*\xi, \eta \rangle} = \langle exp(A^*)exp(\xi), exp(\eta) \rangle$. Hence $exp(A)^* = exp(A^*)$. For P a projection, $exp(P)$ is a projection because it is obviously idempotent. For S an isometry, $exp(S)$ preserves the scalar products of exp-vectors and hence preserves scalar products. If S is also unitary, then $exp(S)$ will be surjective, hence unitary. The converse follows by restricting to K. □

Remark 2.1.14 *Let $K = K_1 \oplus K_2$, $\xi = \xi_1 \oplus \xi_2$. Then there is a unitary $U : \mathcal{F}^s(K) \to \mathcal{F}^s(K_1) \otimes \mathcal{F}^s(K_2)$ s.t.*

$$U(exp(\xi_1 \oplus \xi_2)) = exp(\xi_1) \otimes exp(\xi_2)$$

Proof: $\langle exp(\xi_1 \oplus \xi_2), exp(\eta_1 \oplus \eta_2) \rangle = e^{\langle \xi_1 \oplus \xi_2, \eta_1 \oplus \eta_2 \rangle} = e^{\langle \xi_1, \eta_1 \rangle} e^{\langle \xi_2, \eta_2 \rangle} = \langle exp(\xi_1), exp(\eta_1) \rangle \langle exp(\xi_2), exp(\eta_2) \rangle$ □

Lemma 2.1.15 *Each unitary $U : \mathcal{F}^s(H) \to \mathcal{F}^s(K)$ preserving exp-vectors (up to a constant) is of the form*

$$W(A, \xi, c)exp(\eta) = c\, e^{-\|\xi\|^2/2 - \langle \xi, A\eta \rangle} exp(A\eta + \xi)$$

with unique $c \in S^1 \subseteq \mathbb{C}$, $A : H \to K$ unitary and $\eta \in K$.

Proof: c.f. [Gui 72 Lemma 2.1] (for the case $H = K$) with obvious modifications. □

CONTINUOUS TENSOR PRODUCTS

Theorem 2.1.16 *The group of unitaries preserving exponential vectors in $\mathcal{U}(\mathcal{F}^s(K))$ with the strong operator topology is isomorphic (as a topological group) to the set $\mathcal{U}(K) \times K \times S^1$ where $\mathcal{U}(K)$, K and S^1 are equiped with the strong, norm and usual topology respectively and the composition law*

$$(A, \xi, c)(B, \eta, d) = (AB, \xi + A\eta, cde^{-iIm\langle\xi, A\eta\rangle})$$

Proof: [Gui 72 Thm.2.1] □

B. The Exponential Tensor Decomposition

We refer to Appendix B for direct integrals. Let $\mathcal{K} = \int_I^\oplus \mathcal{K}_t d\mu$ be any direct integral of Hilbert spaces with admissible functions $\mathcal{S} \subseteq \prod_{t \in I} \mathcal{K}_t$ and consider $\mathcal{H} = \mathcal{F}^s(\int_I^\oplus \mathcal{K}_t d\mu)$. For partitions into Borel subsets $P = \{\Delta_1, \ldots, \Delta_k\}$, $Q = \{\Delta_1', \ldots \Delta_l'\}$, $P \leq Q$

$$\varphi_P : \otimes_i exp(f_i) \mapsto exp(\oplus_i f_i) \quad , \quad f_i \in \mathcal{S}|\Delta_i$$

$$\varphi_{PQ} : \otimes_i exp(f_i) \mapsto \otimes_j (exp(\oplus_{\Delta_i \subseteq \Delta_j'} f_i))$$

defines a tensor decomposition over $\mathcal{B}_0(I)$ (even over $\mathcal{B}(I)$) which is denoted by $\bigotimes_I (\mathcal{K}_t, \mu)$.

Definition 2.1.17 *A subset $S \subseteq \int_I^\oplus \mathcal{K}_t d\mu$ is called strongly spanning if $[\{exp(f)|f \in S\}] = \mathcal{F}^s(\int_I^\oplus \mathcal{K}_t d\mu)$.*

Remark 2.1.18 *If μ is a continuous measure, then, for $f, g \in \int_I^\oplus \mathcal{K}_t d\mu$, the map $t \mapsto \chi_{(a,t)} f + \chi_{[t,b)} g \in \int_I^\oplus \mathcal{K}_t d\mu$ is continuous and for a descending sequence of subintervals I_k s.t. $\bigcap_n I_k$ contains at most one point, the sequence $(\chi_{I_k} f)$ tends to zero.*

Proof: μ is continuous iff for each $\varphi \in L^1(I, \mu)$ the map $t \mapsto \int_a^t \varphi(x) d\mu(x)$ is continuous. Thus for any $h \in \int_I^\oplus \mathcal{K}_t d\mu$ and μ continuous $t \mapsto \langle \chi_{(a,t)} f + \chi_{[t,b)} g, h \rangle = \int_a^t \langle f(x), h(x) \rangle d\mu(x) + \int_t^b \langle g(x), h(x) \rangle d\mu(x)$ is continuous. The second statement is obvious. □

The following technical Lemma will be used in Thm.2.2.4, sec.3.3.4 and sec.4.2.4. It is a generalization of a theorem of Arveson [Ar 89 Prop.6.12, Ar 94b Lemma 4.30] which says that subsets which are total and convex are strongly spanning.

Lemma 2.1.19 Let $S \subseteq \int_I^\oplus \mathcal{K}_t d\mu = \mathcal{K}$ be a total (i.e. spanning) subset with this property: For disjoint halfopen subintervals $I_k \subseteq I$ [1] and $s_k \in S$ $\sum_{k=1}^n \chi_{I_k} s_k \in S$. If μ is a continuous measure, then S is strongly spanning.

<u>Proof:</u> If S is as assumed, put $\mathcal{L} := [exp(S)]$. We show inductively that $\mathcal{K}^{\otimes_s^r} \subseteq \mathcal{L}$ for each $r \in \mathbb{N}$. Without loss assume that $I = (0, 1)$. Then

<u>1</u>: $\Omega = exp(0) \in \mathcal{L}$ (obvious because $0 \in S$)

<u>2</u>: Suppose $\mathcal{K}^{\otimes_s^{r'}} \subseteq \mathcal{L}$ for each $r' < r$, $r, r' \in \mathbb{N}$. We need to show that

$$\left(\sum_{i \in J} \alpha_i s^i\right)^{\otimes^r} \in \mathcal{L}$$

for a finite sum with $s^i \in S$ and $\alpha_i \in \mathbb{C}$.

Now for any $m \in \mathbb{N}$, $k \in \{1, \ldots, 2^m\}$ let $J_k^m = [\frac{k-1}{2^m}, \frac{k}{2^m})$, $s_k^i := \chi_{J_k^m} s^i$ (m is omitted). Of course, $\sum_i \alpha_i s^i = \sum_{k,i} \alpha_i s_k^i$, but the tensor powers of this sum can be approximated in terms of the s_k^i, and this is what we mean by letting $m \to \infty$. The following two assertions conclude the proof of <u>2</u>:

Statement A):

$$\left\|\sum \alpha_{j_1} \ldots \alpha_{j_r} \sum_{\sigma \in S_r} s_{k_{\sigma 1}}^{j_{\sigma 1}} \otimes \ldots \otimes s_{k_{\sigma r}}^{j_{\sigma r}} - \left(\sum_{i,k} \alpha_i s_k^i\right)^{\otimes^r}\right\| \to 0$$

for $m \to \infty$ where the big sum is the sum over all subsets $\{(j_1, k_1), \ldots, (j_r, k_r)\} \subseteq J \times \{1, \ldots, 2^m\}$ s.t. $k_1 < \ldots < k_r$.

Statement B) ($\widehat{}$ means omission):

$$\sum_{\sigma \in S_r} s_{k_{\sigma 1}}^{j_{\sigma 1}} \otimes \ldots \otimes s_{k_{\sigma r}}^{j_{\sigma r}} = (s_{k_1}^{j_1} + \ldots + s_{k_r}^{j_r})^{\otimes^r} -$$

$$- \sum_{\{s\} \subseteq \{1,\ldots,r\}} (s_{k_1}^{j_1} + \ldots + \widehat{s_{k_s}^{j_s}} + \ldots + s_{k_r}^{j_r})^{\otimes^r} +$$

$$+ \sum_{\{s_1, s_2\} \subseteq \{1,\ldots,r\}} (s_{k_1}^{j_1} + \ldots + \widehat{s_{k_{s_1}}^{j_{s_1}}} + \ldots + \widehat{s_{k_{s_2}}^{j_{s_2}}} + \ldots + s_{k_r}^{j_r})^{\otimes^r}$$

$$- + \ldots$$

---

[1] It suffices to assume that the endpoints of the I_k lie in a dense subset of I

CONTINUOUS TENSOR PRODUCTS 21

and

$$\left\| \sum \alpha_{j_1} \ldots \alpha_{j_r} \left[(s_{k_1}^{j_1} + \ldots + s_{k_r}^{j_r})^{\otimes \bar{r}} - \sum_{\{s\} \subseteq \{1,\ldots,r\}} (s_{k_1}^{j_1} + \ldots + \widehat{s_{k_s}^{j_s}} + \ldots + s_{k_r}^{j_r})^{\otimes \bar{r}} \right. \right.$$
$$\left. \left. + \sum_{\{s_1, s_2\} \subseteq \{1,\ldots,r\}} (s_{k_1}^{j_1} + \ldots + \widehat{s_{k_{s_1}}^{j_{s_1}}} + \ldots + \widehat{s_{k_{s_2}}^{j_{s_2}}} + \ldots + s_{k_r}^{j_r})^{\otimes \bar{r}} - + \ldots \right] \right\| \to 0$$

for $\bar{r} > r$ and $m \to \infty$.

<u>Proof of A)</u>: Idea: $(\sum_{i,k} \alpha_i s_k^i)^{\otimes r}$ multiplied out contains $|J|^r 2^{mr}$ terms and the big sum in A) consists of $|J|^r 2^m (2^m - 1) \ldots (2^m - r + 1)$ terms so that the remaining ones can be neglected for $m \to \infty$. Let δ be equal to

$$\| \sum_{\exists i \neq j : k_i = k_j} \alpha_{j_1} \ldots \alpha_{j_r} s_{k_1}^{j_1} \otimes \ldots \otimes s_{k_r}^{j_r} \|^2$$

Then δ converges to 0 if $m \to \infty$: Extending s_k^i by 0 to $(0,1)$ consider $\mathsf{s} := (\sum_{k,j} \alpha_j s_k^j) \otimes \ldots \otimes (\sum_{k,j} \alpha_j s_k^j)$ as a function on $(0,1)^r$. Let $D_{m,r} = \{(t_1, \ldots, t_r) \in (0,1)^r \mid \exists k, l : k \neq l \text{ and } |t_l - t_k| \leq \frac{1}{2^{m-1}}\}$. Then $\mu^{\otimes r}(D_{m,r}) \to 0$ for $m \to \infty$ and we have $\delta \leq \frac{1}{\|\mathsf{s}\|^2} \int_{D_{m,r}} (d\mu)^r \langle \mathsf{s}, \mathsf{s} \rangle \xrightarrow{m \to \infty} 0$. Thus we may replace the sum over all different k_i by the sum over all k_i for m sufficiently large.

<u>Proof of B)</u>: The first part is some combinatorics. For the norm assertion we can use a similar argument as in A). We can write the expression in the second part of the statement as $\| \sum \alpha_{j_1} \ldots \alpha_{j_r} \sum s_{l_1}^{i_1} \otimes \ldots \otimes s_{l_{\bar{r}}}^{i_{\bar{r}}} \|$ where the second sum is taken over all couples s.t. $\{(i_1, l_1)), \ldots, (i_{\bar{r}}, l_{\bar{r}})\} = \{(j_1, k_1), \ldots, (j_r, k_r)\}$, $k_1 < \ldots < k_r$ which is also the sum over $\{i_1, \ldots, i_{\bar{r}}\} = \{j_1, \ldots, j_r\}$ and $|\{l_1, \ldots, l_{\bar{r}}\}| = r$. In particular, there must be a repetition in the sequence $l_1, \ldots, l_{\bar{r}}$. Therefore, if $D_{m,\bar{r}} = \{(t_1, \ldots, t_{\bar{r}}) \in (0,1)^{\bar{r}} \mid \exists k, l : k \neq l \text{ and } |t_k - t_l| < \frac{1}{2^{m-1}}\}$, then for fixed $j_1, \ldots, j_r$ we have $\| \sum s_{l_1}^{i_1} \otimes \ldots \otimes s_{l_{\bar{r}}}^{i_{\bar{r}}} \| \leq \| \sum s^{i_1} \otimes \ldots \otimes s^{i_{\bar{r}}} | D_{m,\bar{r}} \|$ where the sum on the r.h.s. is taken over $\{i_1, \ldots, i_{\bar{r}}\} = \{j_1, \ldots, j_r\}$. For $\bar{r}$ fixed and $m \to \infty$ it converges to 0 and for m fixed and $\bar{r} \to \infty$ the norm of the sum is bounded by $k c^{\bar{r}}$ with some constants $k, c > 0$

<u>Proof of the induction assertion:</u> By induction assumption, we only have to

prove that if P_r is the projection in $\mathcal{F}^s$ into the $\bar{r}$-particle components $\bar{r} \geq r$, then

$$P_r\left(\sum \alpha_{j_1}\ldots\alpha_{j_r}\sqrt{r!}\left[exp(s_{k_1}^{j_1}+\ldots+s_{k_r}^{j_r})-\right.\right.$$

$$-\sum_{\{s\}\subseteq\{1,\ldots,r\}} exp(s_{k_1}^{j_1}+\ldots+\widehat{s_{k_s}^{j_s}}+\ldots+s_{k_r}^{j_r})$$

$$\left.\left.+\sum_{\{s_1,s_2\}\subseteq\{1,\ldots,r\}} exp(s_{k_1}^{j_1}+\ldots+\widehat{s_{k_{s_1}}^{j_{s_1}}}+\ldots+\widehat{s_{k_{s_2}}^{j_{s_2}}}+\ldots+s_{k_r}^{j_r})-+\ldots\right]\right)$$

converges to $\left(\sum_i \alpha_i s^i\right)^{\otimes^r}$ for $m \to \infty$.

A) and the first part of B) implies that the r-th component converges to $(\sum_i \alpha_i s^i)^{\otimes^r}$. The second part of B) implies that the higher order terms tend to zero for any fixed $\bar{r} > r$ and for m fixed the norms of the components are bounded by $\frac{k_1 c_1^{\bar{r}}}{\sqrt{\bar{r}!}}$ for some constants $k_1, c_1 > 0$. But $\sum_{\bar{r}>r_0} k_1^2(\frac{c_1^{\bar{r}}}{\sqrt{\bar{r}!}})^2$ converges to 0 for $r_0 \to \infty$ which concludes the proof. $\square$

Proposition 2.1.20 *The exponential tensor decomposition is continuous if μ is a continuous measure.*

<u>Proof:</u> 1.) Weak continuity: By 2.1.12, we only need $S \subseteq \int_I^\oplus \mathcal{K}_t d\mu$ countable strongly spanning s.t. for $s, s' \in S$, $s \neq s'$ implies $s_t \neq s'_t$. Such S exist (polynomials in each component of $\int_I^\oplus \mathcal{K}_t d\mu = \oplus L^2(\Lambda_n, \mathbb{C}^n)$ multiplied with a weight function in case μ is unbounded with coefficients in $\mathbb{Q}(i)$). For such an $S = \{s^i | i \in \mathbb{N}\}$ the Gram-Schmidt procedure is never singular and we get onb's

$$e_n(t) = \sum_{k=0}^n \lambda_{kn}(t) exp(s_t^k) \text{ of } \mathcal{H}_t$$

$$f_m(t) = \sum_{l=0}^m \mu_{lm}(t) exp(s^{l,t}) \text{ of } \mathcal{H}^t$$

with continuous coefficients. For continuity we show: Let $M_t := \mathcal{B}(\mathcal{H}_t) \otimes \mathbf{1}$. Then $(\sqcup_t M_t)'' = \mathcal{B}(\mathcal{H}) =: M$ (the rest follows by taking commutants). But for $f, g, h, k \in \int_I^\oplus \mathcal{K}_t d\mu$

$$\langle exp(h), (\Theta_{exp(f_t),exp(g_t)} \otimes \mathbf{1}) exp(k)\rangle = e^{\langle h_t, f_t\rangle + \langle g_t, k_t\rangle + \langle h^t, k^t\rangle}$$

Because exponential vectors are total (2.1.12.(ii)) $t \mapsto \Theta_{exp(f_t),exp(g_t)} \otimes \mathbf{1}$ converges weakly to $\Theta_{exp(f),exp(g)}$. Hence $(\bigcup_t M_t)''$ contains the compacts. □

Remark 2.1.21 (i) Each exp-vector is factorizable in $\bigotimes_I(\mathcal{K}_t,\mu)$. Conversely, it will be seen in 2.2.4 that each vector factorizable w.r.t. $\mathcal{B}_0(I)$ is (a multiple of) an exp-vector ([ArWo 66]).

(ii) For $\int_I^\oplus \mathcal{K}_t d\mu = L^2(I,\mathbb{C}^n)$, we denote $\bigotimes_I(\mathcal{K}_t,\mu)$ by $T_n(I)$.

C. Rearranged Exponential Tensor Decompositions

We use the same simple idea as in 2.1.3.C to exhibit a one parameter family of pairwise inequivalent type III tensor decompositions. First we have the following observation due to [ArWo 66]:

Lemma 2.1.22 For $I = \bigcup_{k \in \mathbb{N}} \Lambda_k$ a countable partition into Borel subsets of positive Lebesgue measure, let $\mathcal{H} = \mathcal{F}^s(L^2(I,\mathbb{C}^n))$ and $\Omega_k = exp(0|\Lambda_k)$ the vacuum in $\mathcal{F}^s(L^2(\Lambda_k,\mathbb{C}^n))$. Then $\mathcal{F}^s(L^2(I,\mathbb{C}^n)) \cong \bigotimes_k (\mathcal{F}^s(L^2(\Lambda_k,\mathbb{C}^n)), \Omega_k)$ where $exp(f)$ is mapped to $\otimes_k exp(f|\Lambda_k)$.

Proof: Clear. □

Let $I := (0,1) = \bigcup_n I_n$, $I_n = [2^{-n-1}, 2^{-n})$, $H_n = \mathcal{F}^s(L^2(I_n, \mathbb{C}^2)) = H_{(n,0)} \otimes H_{(n,1)}$, $H_{(n,i)} = \mathcal{F}^s(L^2(I_n))$, $e_1^{(n,i)} = exp(0)$, $e_2^{(n,i)} = 2^{n+1}\chi_{I_n} \otimes \chi_{I_n}$ ($\chi_{I_n} \otimes \chi_{I_n}$ in the two particle component) both in $H_{(n,i)}$. Put

$$\Omega_n^\lambda := \lambda^{\frac{1}{2}}(\lambda+1)^{-\frac{1}{2}} e_1^{(n,0)} \otimes e_1^{(n,1)} + (1+\lambda)^{-\frac{1}{2}} e_2^{(n,0)} \otimes e_2^{(n,1)}$$

Let $\mathcal{H}^\lambda := \bigotimes_n (H_n, \Omega_n^\lambda)$ be the infinite tensor product and define a rearrangement map

$$\sigma : I \to I \dot\cup I = I \times \{0,1\}$$

by

$$\sigma(t) = \begin{cases} (2^{-n-1} + 2(t - 2^{-n-1}), 0) & \text{if } t \in [2^{-n-1}, 2^{-n-1} + 2^{-n-2}) \\ (2^{-n-1} + 2(t - 2^{-n-1} - 2^{-n-2}), 1) & \text{if } t \in [2^{-n-1} + 2^{-n-2}, 2^{-n}) \end{cases}$$

For $J = [x,y] \subseteq (0,1)$ and $x > 0$, define

$$\mathcal{H}^\lambda{}_J = \mathcal{F}^s(L^2(\sigma^{-1}(J)))$$

If $J = (0, y) \subseteq I$, let n_0 be s.t. $2^{-n_0} < y$ and put

$$\mathcal{H}^\lambda{}_J = (\bigotimes_{n \geq n_0} (H_n, \Omega_n^\lambda)) \otimes \mathcal{H}^\lambda{}_{[2^{-n_0}, y)}$$

$\mathcal{H}^\lambda{}_J$ may be considered as a subspace of $\mathcal{H}^\lambda$ using $\Omega^\lambda = \otimes \Omega_n^\lambda$ and does not depend on the choice of n_0 in case $J = (0, y)$. We obtain a tensor decomposition over $\mathcal{B}_0(I)$:

For any partition of $I = (0, 1)$ into subintervals $J_1 = (0, t_1), J_2 = [t_1, t_2), \ldots, J_m = [t_{m-1}, 1)$, let $n_0 \in \mathbb{N}$ be s.t. $2^{-n_0} < t_1$ and put $f_l := (f \circ \sigma)|J_l$ for any $f \in L^2[2^{-n_0}, 1)$. Then for $\xi \in \mathcal{H}^\lambda{}_{(0, 2^{-n_0})}$, we have $\xi_1 = \xi \otimes exp(f_1) \in \mathcal{H}^\lambda{}_{(0, t_1]}$, $\xi_l := exp(f_l) \in \mathcal{H}^\lambda{}_{J_l}$ and we can define for partitions $P, Q \geq \{J_1, \ldots, J_m\}$, $P \leq Q$:

$$\varphi_P(\otimes \xi_l) = \otimes_{\Delta \in P}(\otimes_{J_l \subseteq \Delta} \xi_l) \quad \varphi_{PQ}(\otimes \xi_l) = \otimes_{\Delta' \in Q}(\otimes_{J_l \subseteq \Delta'} \xi_l)$$

For the continuity condition, note that in an infinite tensor product $\mathcal{H} = \bigotimes_{n \in \mathbb{N}}(\mathcal{H}_n, \Omega_n)$ the subalgebras $M_N = \mathcal{B}(H_0 \otimes \ldots \otimes H_N) \otimes \mathbf{1}$ have the property $(\bigcup_N M_N)'' = \mathcal{B}(\mathcal{H})$. The proof is the same as in 2.1.20 replacing exp-vectors by product vectors. (Weak continuity is also clear)

Lemma 2.1.23 *Suppose $\xi \in \mathcal{H}^\lambda$ is a factorizable unit vector. Then there exists $f : (0, 1) \to \mathbb{C}$ Lebesgue measurable and a family of unit vectors $\xi_\varepsilon \in \mathcal{H}^\lambda{}_\varepsilon = \mathcal{H}^\lambda{}_{(0, \varepsilon)}$ s.t. $f^\varepsilon := f|(\varepsilon, 1) \in L^2(\varepsilon, 1)$ and $\xi = \xi_\varepsilon \otimes e^{-\|f^\varepsilon\|^2/2} exp(f^\varepsilon)$, $\varepsilon \in (0, 1)$.*

Proof: (We need Thm.2.2.4) For $1 > \varepsilon > 0$, we have $\xi = \xi_\varepsilon \otimes \xi^\varepsilon$, $\|\xi_\varepsilon\| = \|\xi^\varepsilon\| = \|\xi\| = 1$, $\xi_\varepsilon \in \mathcal{H}^\lambda{}_\varepsilon$, $\xi^\varepsilon \in \mathcal{H}^{\lambda, \varepsilon} = \mathcal{F}(L^2(\varepsilon, 1))$. ξ^ε is again factorizable and by Thm.2.2.4 of the form $c(\varepsilon) e^{-\|f(\varepsilon)\|^2/2} exp(f(\varepsilon))$, $|c(\varepsilon)| = 1$ for some $f(\varepsilon) \in L^2(\varepsilon, 1)$. But we also have $\xi^{\varepsilon'} = \xi_{[\varepsilon', \varepsilon)} \otimes \xi^\varepsilon$, $0 < \varepsilon' < \varepsilon$ for some $\xi_{[\varepsilon', \varepsilon)} = c(\varepsilon', \varepsilon) e^{-\|f(\varepsilon', \varepsilon)\|^2/2} exp(f(\varepsilon', \varepsilon))$, $|c(\varepsilon', \varepsilon)| = 1$. Finally $c(\varepsilon') e^{-\|f(\varepsilon')\|^2/2} exp(f(\varepsilon')) = c(\varepsilon', \varepsilon) c(\varepsilon) e^{-\|f(\varepsilon)\|^2/2 - \|f(\varepsilon', \varepsilon)\|^2/2} exp(f(\varepsilon', \varepsilon)) \otimes exp(f(\varepsilon))$. Hence $f(\varepsilon') = f(\varepsilon', \varepsilon) \oplus f(\varepsilon)$ by linear independence of exp-vectors. □

Lemma 2.1.24 *For each $n \in \mathbb{N}$ and $f \in L^2(I_n, \mathbb{C}^2)$ we have $dist(\mathbb{C}exp(f), \Omega^\lambda) > k(\lambda) > 0$ where $k(\lambda)$ only depends on $\lambda \in (0, 1)$.*

Proof: For $f = f_1 \oplus f_2$, $f_i \in L^2(I_n, \mathbb{C})$, we have (omitting n)

$$dist(\mathbb{C}exp(f), \Omega^\lambda)^2 = 1 - e^{-\|f_1\|^2 - \|f_2\|^2} |\langle exp(f_1) \otimes exp(f_2), \Omega^\lambda \rangle|^2$$

$$= 1 - e^{-(\|f_1\|^2 + \|f_2\|^2)} \left| \lambda^{\frac{1}{2}} (1+\lambda)^{-\frac{1}{2}} + (1+\lambda)^{-\frac{1}{2}} \right.$$
$$\left. \frac{1}{2} \langle f_1, 2^{(n+1)/2} \chi_{I_n} \rangle^2 \langle f_2, 2^{(n+1)/2} \chi_{I_n} \rangle^2 \right|^2$$

$$\geq 1 - e^{-(\|f_1\|^2 + \|f_2\|^2)} \left| \lambda^{\frac{1}{2}} (1+\lambda)^{-\frac{1}{2}} + (1+\lambda)^{-\frac{1}{2}} \frac{1}{2} \|f_1\|^2 \|f_2\|^2 \right|^2$$

by the Cauchy-Schwarz inequality.

But a simple computation shows that for $x, y \geq 0$ and $\lambda \in (0,1)$ we have

$$e^{-\frac{x+y}{2}} (1+\lambda)^{-\frac{1}{2}} (\lambda^{\frac{1}{2}} + \frac{1}{2} xy) \leq c(\lambda) < 1$$

for some suitable constant $c(\lambda)$. So $k(\lambda) = \sqrt{1 - c(\lambda)^2}$ will do. $\square$

Theorem 2.1.25 $(\mathcal{H}^\lambda, \mathcal{B}_0(I))$ *is a family of inequivalent continuous tensor decompositions of type III.*

Proof: Let ξ be a factorizable unit vector in $\mathcal{H}^\lambda$ and $f : (0,1) \to \mathbb{C}$ as in 2.1.23. By restricting f to subintervals, we get $\xi_n \in H_n$ unit vectors which are multiples of exponential vectors s.t. $\xi = \otimes \xi_n$. But this is impossible by 2.1.24. The inequivalence for different λ follows as in sec.2.1.3.C using Lemma 2.1.9.(ii). $\square$

As in the discrete case there appears to be no way to extend this tensor decomposition to a complete Boolean algebra. In fact it is likely that each (separable) continuous tensor decomposition over the Borel sets is exponential.

It seems possible to define an analogue of Araki and Woods' asymptotic ratio set $r_\xi \subseteq [0, \infty)$ for vectors ξ in tensor decompositions (c.f. [ArWo 68]) in order to obtain invariants for nonexponential tensor decompositions and product systems.

2.2 The Generalized Araki-Woods Theorem

In [ArWo 66] (c.f. [Gui 72]) Araki and Woods proved the following remarkable theorem saying in our terminology:

Let $T = (\mathcal{H}, \mathcal{B})$ be a tensor decomposition of type I over $\mathcal{B}$, complete and nonatomic with the following properties: There is a factorizable unit vector $\Omega \in \mathcal{F}$ and a family $(\Omega_\Delta)_{\Delta \in \mathcal{B}}$ of unit vectors s.t. for each finite partition P of $1 \in \mathcal{B}$ and each countable partition P_Δ of any given $\Delta \in \mathcal{B}$, we have $\Omega = \bigotimes_{\Delta \in P} \Omega_\Delta$ and $(\mathcal{H}_\Delta, \Omega_\Delta) \cong \bigotimes_{\Delta_i \in P_\Delta}(\mathcal{H}_{\Delta_i}, \Omega_{\Delta_i})$ via the infinite tensor product map. Then T is exponential.

Because of noncompleteness and the fact that only finite partitions are considered we can't apply this result to tensor decompositions over $\mathcal{B}_0(\bar{I})$ in which we are interested because they are the ones occuring in connection with product systems.

We are going to show that an Araki-Woods-theorem holds also in this situation. Our proof is more in the spirit of the spectral theorem in the direct integral version and motivated by Arveson's work on divisible product systems [Ar 89].

2.2.1 An Araki-Woods Theorem for $\mathcal{B}_0(I)$

Lemma 2.2.1 *Let $T = (\mathcal{H}, \mathcal{B}_0(I))$ be continuous of type I and ξ, η nonzero factorizable vectors. Then $\langle \xi, \eta \rangle \neq 0$.*

<u>Proof</u>: We may assume ξ, η to be unit vectors. Let $t_0 \in I$. Then by Lemma 2.1.5.

$$(*) \quad \bigcap_{\varepsilon > 0} M_{[t_0 - \varepsilon, t_0 + \varepsilon]} = \mathbb{C}\mathbf{1}$$

where $M_{[t_0 - \varepsilon, t_0 + \varepsilon]} = \mathcal{B}(\mathcal{H}_{[t_0 - \varepsilon, t_0 + \varepsilon]}) \otimes \mathbf{1}$.

Let $\omega_\xi = \langle \xi, \cdot\, \xi \rangle$ and $\omega_\eta = \langle \eta, \cdot\, \eta \rangle$. We have $\omega_\xi | M_{[t_0 - \varepsilon, t_0 + \varepsilon]} = \langle \xi_{[t_0 - \varepsilon, t_0 + \varepsilon]}, \cdot\, \xi_{[t_0 - \varepsilon, t_0 + \varepsilon]} \rangle$ where $\xi = \xi_{t_0 - \varepsilon} \otimes \xi_{[t_0 - \varepsilon, t_0 + \varepsilon]} \otimes \xi^{t_0 + \varepsilon}$ are all unit vectors and unique up to a scalar of absolute value 1 (the same for η). Now

$$|\langle \xi_{[t_0 - \varepsilon, t_0 + \varepsilon]}, \eta_{[t_0 - \varepsilon, t_0 + \varepsilon]} \rangle|^2 = 1 - \tfrac{1}{4}\|(\omega_\xi - \omega_\eta)|M_{[t_0 - \varepsilon, t_0 + \varepsilon]}\|^2$$

Hence by $(*)$ we get (using normality of the states)

$$\lim_{\varepsilon \to 0} |\langle \xi_{[t_0-\varepsilon,t_0+\varepsilon)}, \eta_{[t_0-\varepsilon,t_0+\varepsilon)} \rangle| = 1$$

The same argument works for the possibly infinite boundary points by continuity at those points. Using compactness of $\bar{I} = [a,b]$, we obtain a finite sequence $a < t_1 < \ldots < t_n < b$ s.t. $|\langle \xi_{(a,t_1)}, \eta_{(a,t_1)} \rangle|$, $|\langle \xi_{[t_1,t_2)}, \eta_{[t_1,t_2)} \rangle|$, $\ldots$, $|\langle \xi_{[t_n,b)}, \eta_{[t_n,b)} \rangle|$ are all nonzero. But the product of them is $|\langle \xi, \eta \rangle|$. $\square$

We now need a multiplicative version of what is sometimes called a premeasure. A finite premeasure μ on $\mathcal{B}_0(I)$ is a countably additive complex set function in the sense that $\mu(\Delta) = \sum_{i=0}^{\infty} \mu(\Delta_i)$ if $\bigcup_i \Delta_i = \Delta \in \mathcal{B}_0(I)$. μ has a unique extension to the Borel sets because its distribution is of bounded variation.

Suppose we have a function $M : \mathcal{B}_0(I) \to \mathbb{C} \setminus \{0\}$ with the following properties:

(i) For each sequence (I_k) of disjoint halfopen subintervals $I_k \subseteq I$ s.t. $\bigcup_{k=1}^{\infty} I_k \in \mathcal{B}_0(I)$, the infinite product $\prod_k M(I_k)$ is convergent and $\prod_{k=0}^{\infty} M(I_k) = M(\bigcup_{k=0}^{\infty} I_k)$.

(ii) For each descending sequence $(I_k \in \mathcal{B}_0(I))$ of halfopen subintervals s.t. $\bigcap_k I_k$ contains at most one point, $M(I_k) \overset{k \to \infty}{\longrightarrow} 1$.

M is a multiplicative version of a continuous premeasure. The following is a classical fact about multiplicative measures, i.e. functions as above defined on σ-algebras (c.f. [PaSc 72], [Gui 72]). We have adapted the proof from [PaSc 72] to the case of $\mathcal{B}_0(I)$.

Proposition 2.2.2 *Suppose M fulfils the two conditions above. Then there exists a finite continuous complex measure μ on the Borel sets $\mathcal{B}(I)$ s.t. $M(\Delta) = e^{\mu(\Delta)}$ for each $\Delta \in \mathcal{B}_0(I)$.*

Proof: We indicate the steps:

(i) Condition (i) holds if we replace intervals by elements in $\mathcal{B}_0(I)$.

(ii) $\Delta \mapsto \log|M(\Delta)|$ defines a finite premeasure on $\mathcal{B}_0(I)$ and has a unique extension to the Borel sets. In particular $\beta := \sup_{\Delta \in \mathcal{B}_0(I)} |M(\Delta)|$ is finite. (c.f.[Bau 78 5.7])

(iii) Let $\alpha(\Delta) := \sup_{\Delta' \subseteq \Delta} |1 - M(\Delta')|$ be the oscillation of M which is finite for each Δ by (ii). If $\bigcup_i \Delta_i = \Delta \in \mathcal{B}_0(I)$, then

$$\alpha(\bigcup_i \Delta_i) \leq \beta \sum_{i=0}^{\infty} \alpha(\Delta_i)$$

Proof : For $\varepsilon > 0$ take $D \in \mathcal{B}_0(I), D \subset \Delta$ s.t. $|M(D) - 1| \geq \alpha(\Delta) - \varepsilon$. $D \cap \Delta_i$ as well as $D \cap \Delta$ lie in $\mathcal{B}_0(I)$ and $|M(D) - 1| = |\prod_{n=0}^{\infty} M(D \cap \Delta_i) - 1| \leq \sum_{n=0}^{\infty} |M(\bigcup_{k=n+1}^{\infty} D \cap \Delta_k)| |M(D \cap \Delta_n) - 1| \leq \beta \sum_{n=0}^{\infty} \alpha(\Delta_n)$.

(iv) $\alpha([t - \varepsilon, t + \varepsilon)) \xrightarrow{\varepsilon \to 0} 0$ for $t \in I$ and $\alpha((a, a + \varepsilon)), \alpha((b - \varepsilon, b)) \xrightarrow{\varepsilon \to 0} 0$

Proof : For simplicity let $t = 0$ and show only $\alpha((0, t_n)) \to 0, t_n \searrow 0$. Assume $\alpha((0, t_n)) > c > 0$ for all $n \in \mathbb{N}$. There is a sequence $D_k \subseteq (0, t_k), D_k \in \mathcal{B}_0(I)$ s.t. $|M(D_k) - 1| \geq c$. Put $D_k^l := D_k \cap [t_{k+l+1}, t_{k+l})$. Then $M(D_k) = \prod_l M(D_k^l)$ by condition (i). For each k there is l_k s.t. $|M(D_k \cap [t_{k+l_k+1}, t_k)) - 1| > c/2$. Thus we find a sequence n_k s.t. $n_{k+1} > n_k + l_{n_k}$ for each $k \in \mathbb{N}$ and $R_k := D_k \cap [t_{n_{k+1}}, t_{n_k})$ is a disjoint sequence in $\mathcal{B}_0(I)$ s.t. $\prod_k M(R_k)$ does not converge. Now the sequence (R_k) is a decending sequence of elements in $\mathcal{B}_0(I)$ and we can find a sequence in $\mathcal{B}_0(I)$ s.t. its union lies in $\mathcal{B}_0(I)$ and (R_k) being a subsequence. This is a contradiction.

(v) There exists a unique finite premeasure μ on $\mathcal{B}_0(I)$ s.t. $M(\Delta) = e^{\mu(\Delta)}$ for Δ in $\mathcal{B}_0(I)$ (cover I with open subsets U s.t. $\alpha(U)$ are sufficiently small. For the boundary points take halfopen intervalls. Then taking the logarithms of $M(\Delta), \Delta \subset U$ is well defined and gives a premeasure on U which extends to I (compare with [PaSc 72 Thm.4.7])).

(vi) μ has a unique extension to the Borel subsets of I. □

Proposition 2.2.3 *Let $\xi \in \mathcal{F}$ be a factorizable unit vector and consider the set of factorizable vectors $\mathcal{F}_\xi := \{\eta \in \mathcal{F} \,|\, \langle \eta, \xi \rangle = 1\}$. Let $\xi_t \in \mathcal{H}_t$, $\xi^t \in \mathcal{H}^t$ be any choice of unit vectors s.t. $\xi = \xi_t \otimes \xi^t$ for all $t \in I$. For $\eta \in \mathcal{F}_\xi$ let $\eta_t \in \mathcal{H}_t$ and $\eta^t \in \mathcal{H}^t$ be the unique families s.t. $\eta = \eta_t \otimes \eta^t$ and $\langle \eta_t, \xi_t \rangle = 1$ for all $t \in I$. Then for any $\eta^1, \eta^2 \in \mathcal{F}_\xi$ there exists a finite continuous complex measure $\mu_{1,2}$ on I s.t. for each $t \in I$, $\langle \xi_t, \eta_t \rangle = e^{\mu_{1,2}((a,t))}$.*

Proof: For $[s,t) \subseteq I$ define $\xi_{[s,t)}$ by the requirement $\xi_t = \xi_s \otimes \xi_{[s,t)}$ and for $\eta \in \mathcal{F}_\xi$ define $\eta_{[s,t)}$ by $\eta_t = \eta_s \otimes \eta_{[s,t)}$. Then $\langle \xi_{[s,t)}, \eta_{[s,t)} \rangle = 1$. Let (I_k) be a sequence of disjoint halfopen subintervals s.t. $\bigcup_k I_k = J \subseteq I$ is halfopen (or (a,t)). There is a natural embedding of the infinite tensor product $\bigotimes(\mathcal{H}_{I_k}, \xi_{I_k})$ into $\mathcal{H}_J$ mapping any $\zeta \in \mathcal{H}_{I_0} \otimes \ldots \otimes \mathcal{H}_{I_N}$ to $\zeta \otimes \xi_{J \cap I_0^c \cap \ldots \cap I_N^c}$. If $J = [s,t)$ we can find a partition of (I_k) into increasing subsequences $(I_{k_{l_n}})$ s.t. $\bigcup_n I_{k_{l_n}} = J_l$ form an increasing sequence of halfopen intervals s.t. $\bigcup_l J_l = J$. Using this and a similar statement for $J = (a,t)$, the continuity requirement shows that

$$\left(\bigcup_{\substack{F \subseteq \mathbb{N} \\ \text{finite}}} \mathcal{B}(\otimes_{k \in F} \mathcal{H}_{I_k}) \otimes \mathbf{1} \right)'' = \mathcal{B}(\mathcal{H}_J) \otimes \mathbf{1}$$

which implies that the embedding is surjective. Now $\langle \eta_{I_k}, \xi_{I_k} \rangle = 1$ for each $k \in \mathbb{N}$ and thus $\sum_k |\langle \eta_{I_k}, \xi_{I_k} \rangle - 1| = 0$ is certainly convergent. To show convergence of $\prod_k \|\eta_{I_k}\|$ remark that for any finite $F \subseteq \mathbb{N}$ we have $\prod_{k \in F} \|\eta_{I_k}\| = \|\langle \eta, \cdot \xi\rangle | \mathcal{B}(\otimes_{k \in F} \mathcal{H}_{I_k}) \otimes \mathbf{1}\|$ and again the continuity of the t.d. implies that $\prod_k \|\eta_{I_k}\|$ converges to $\|\eta\|$. Thus (η_{I_k}) defines a product vector $\otimes \eta_{I_k}$ and for any two $\eta^1, \eta^2 \in \mathcal{F}_\xi$ it follows that $\langle \eta^1, \eta^2 \rangle = \prod_k \langle \eta^1_{I_k}, \eta^2_{I_k} \rangle$ is convergent. $M_{1,2}(J) := \langle \eta^1_J, \eta^2_J \rangle$ extends therefore to a function $M_{1,2} : \mathcal{B}_0(I) \to \mathbb{C} \setminus \{0\}$ with property (i) above. To check property (ii) we show that $\langle \eta^1_{[s,t)}, \eta^2_{[s,t)} \rangle \to 1$ provided $t \to s$. By continuity, $\|\eta_{[s,t)}\| = \|\langle \eta, \cdot \xi\rangle | M_{[s,t)}\| \to 1$ for $t \to s$. Now we can use an argument of Arveson [Ar 94b 7.5]:

$$\begin{aligned} |1 - \langle \eta^1_{[s,t)}, \eta^2_{[s,t)} \rangle| &= |\langle \xi_{[s,t)} - \eta^1_{[s,t)}, \xi_{[s,t)} - \eta^2_{[s,t)} \rangle| \\ &\leq \|\xi_{[s,t)} - \eta^1_{[s,t)}\| \|\xi_{[s,t)} - \eta^2_{[s,t)}\| \\ &= |1 - \langle \eta^1_{[s,t)}, \eta^1_{[s,t)} \rangle|^{1/2} |1 - \langle \eta^2_{[s,t)}, \eta^2_{[s,t)} \rangle|^{1/2} \end{aligned}$$

where the last equality follows from the the first one. It also follows that

$\langle \eta_t^1, \eta_t^2 \rangle \xrightarrow{t \to a} 1$. Similarly $\langle \eta^{1,t}, \eta^{2,t} \rangle \to 1$ for $t \to b$ and thus $\langle \eta_t^1, \eta_t^2 \rangle \xrightarrow{t \to b} \langle \eta^1, \eta^2 \rangle$. Hence $M_{1,2}$ fulfils also property (ii) and by Prop.2.2.2 there are continuous measures $\mu_{1,2}$ on I s.t. $\langle \eta_t^1, \eta_t^2 \rangle = e^{\mu_{1,2}((a,t))}$ for each $t \in I$ and $\mu_{1,2}((a,t)) \to 0$ for $t \to a$ which must be bounded. □

Theorem 2.2.4 *Let $T = (\mathcal{H}, \mathcal{B}_0(I))$ be a continuous tensor decomposition of type I. Then there exists a continuous measure μ on I and a direct integral of Hilbert spaces $\int_I^\oplus \mathcal{K}_t d\mu$ s.t.*

$$T \cong \bigotimes_I (\mathcal{K}_t, \mu)$$

and the factorizable vectors are mapped to the exponential vectors.

<u>Proof:</u> Assume $I = (0,1)$ and fix any factorizable unit vector $s^o \in \mathcal{F}$ together with decompositons $s^o = s_t^o \otimes s^{o,t}$ into unit vectors for any t. Let $S = \{s^i | i \in \mathbb{N}^*\}$ be a countable dense subset of factorizable vectors in $\mathcal{F}_{s^o} = \{s \in \mathcal{F} | \langle s, s^o \rangle = 1\}$ and fix as before $s^i = s_t^i \otimes s^{i,t}$ by requirering $\langle s_t^i, s_t^o \rangle = 1$. By the foregoing proposition there are continuous measures μ_{ij} on I with $\langle s_t^i, s_t^j \rangle = e^{\mu_{ij}((0,t))}$ where $\mu_{oi} = \mu_{io} = \mu_{oo} = 0$. We may assume now that S has the following property: For a subinterval $[x,y) \subseteq I$ and $s \in S$, let $s_{[x,y)}$ be given by $s_y = s_x \otimes s_{[x,y)}$. For each partition $I = \bigcup_{k=1}^{2^m} I_k$ with $I_k := [\frac{k-1}{2^m}, \frac{k}{2^m})$ $I_1 = (0, \frac{1}{2^m})$ and $s^1, \ldots, s^{2^m} \in S$, we require $s_{I_1}^1 \otimes \ldots \otimes s_{I_{2^m}}^{2^m} \in S$ (We can always add these elements to S and everything remains countable).

If $\Lambda \subseteq I$ is a Borel set, put

$$\mu(\Lambda) := \sum_{\mu_{ij} \neq 0} \frac{\mu_{ij}(\Lambda)}{2^{i+j}|\mu_{ij}|(I)}$$

Then μ is again continuous s.t. $\mu_{ij} \ll \mu$. By the Radon-Nikodym theorem, there is a Borel function ρ_{ij} s.t.

$$\mu_{ij}(\Lambda) = \int_\Lambda \rho_{ij}(t) d\mu(t)$$

Because $e^{\mu_{ij}(\Delta)} = \langle s_\Delta^i, s_\Delta^j \rangle$ for $\Delta \in \mathcal{B}_0(I)$ the matrices given by $\mu_{ij}(\Delta)$ and $\rho_{ij}(t) = \lim_{\varepsilon \to 0} \frac{1}{\varepsilon} \mu_{ij}([t, t+\varepsilon))$ are conditionally positive definite and $\rho_{ij}(t) = \rho_{ij}(t) - \rho_{oj}(t) - \rho_{io}(t) + \rho_{oo}(t)$ is positive definite (Appendix Prop.15). By

Appendix B.2 the family $t \mapsto \rho_{ij}(t)$ defines a direct integral $\int_I^\oplus \mathcal{K}_t d\mu$ with $\mathcal{K}_t$ separable. If $J_t : S \to \mathcal{K}_t$ is the identification map (i.e. the image of s in the completion of $(\mathbb{C}S, \langle \cdot, \cdot \rangle_t))$, then we have the following map $\varphi : S \to \mathcal{F}^s(\int_I^\oplus \mathcal{K}_t d\mu)$ sending

$$s^i \mapsto exp(I \ni t \mapsto J_t s^i)$$

and also φ_Δ sending

$$s^i_\Delta \mapsto exp(\Delta \ni t \mapsto J_t s^i)$$

Because φ and φ_Δ are scalar product preserving they extend to embeddings of $\mathcal{H}$ and $\mathcal{H}_\Delta$. We only need to check surjectivity: But because the image of s^o_Δ under φ_Δ is $exp(0)$ our assumption on S implies that $\varphi(S)$ is a set of exp-vectors $\{exp(f^i) | i \in \mathbb{N}\}$ s.t. $\{f^i | i \in \mathbb{N}\}$ fulfils the condition in Lemma 2.1.19 and thus is strongly spanning. □

Now we obtain as in [Gui 72]:

Proposition 2.2.5 *Let* $\Phi : \bigotimes_I (\mathcal{K}_t, \mu) \to \bigotimes_I (\mathcal{K}_t, \mu)$ *be factorizable (c.f.2.1.1.(ii)) and unitary. Then there exists a unique* $f \in \int_I^\oplus \mathcal{K}_t d\mu$, *a diagonal unitary* A *and* $c \in S^1$ *s.t.*

$$\Phi(exp(h)) = c\, e^{-\|f\|^2/2 - \langle f, Ah \rangle} exp(Ah + f)$$

The group $\mathcal{G}$ of factorizable unitaries is strongly closed.

Proof: By 2.2.4. we know that Φ preserves exponential vectors hence by 2.1.15.(i) is of the form $W(A, f, c)$, $A \in \mathcal{B}(\int_I^\oplus \mathcal{K}_t d\mu)$ unitary, $h \in \int_I^\oplus \mathcal{K}_t d\mu$, $c \in S^1$. We only have to check that A is diagonal and this can be done exactly as in [Gui 72 Thm.5.2]. $\mathcal{G}$ is strongly closed because it is topologically isomorphic to $S^1 \times \mathcal{D}_u \times \int_I^\oplus \mathcal{K}_t d\mu$ ($\mathcal{D}_u$ the diagonal unitaries) with the multiplication

$$(A, f, c)(B, g, d) = (AB, f + Ag, cd\, e^{-iIm\langle f, Ag \rangle})$$

and this group is complete. □

We note for later use:

Remark 2.2.6 *Recall that a conjugation in a Hilbert K space is a conjugate linear map j s.t. $j^2 = 1$. For any conjugation j also $j^{\otimes^k}$ is a conjugation on $K^{\otimes^k}$ and it is obvious how to define $\hat{J} := exp(j)$ on $\mathcal{F}^s(K)$. If $j(t)$ is a conjugation of $\mathcal{K}_t$ for μ almost all $t \in I$, then $j := \int^\oplus j(t) d\mu(t)$ is a conjugation of $\mathcal{F}^s(\int_I^\oplus \mathcal{K}_t d\mu)$ which is factorizable in the obvious sense. Each conjugate unitary factorizable map is of the form $\hat{J}\Phi$ for Φ factorizable unitary.*

Chapter 3

Algebras Associated to Continuous Tensor Products

To a tensor decomposition T over $\mathcal{B}_0(I)$ we can associate a continuous analogue of a UHF-algebra denoted by $\mathcal{A}(T)$ (which is nonsimple however). It is only further treated in case T is of type I. Throughout we mean by I the interval (a,b) and often assume $a=0$, $b=1$. In principle there are no differences between the finite and infinite interval situation.

3.1 Definition of $L^1(T)$ and $\mathcal{A}(T)$

For a tensor decomposition $T = (\mathcal{H}, \mathcal{B}_0(I))$ over $\mathcal{B}_0(I)$, consider the following Borel subspace of $I \times \mathcal{B}(\mathcal{H})$ with the Borel structure given by the strong topology:

$$\mathbb{K} = \bigcup_{t \in I} \{t\} \times (\mathcal{K}(\mathcal{H}_t) \otimes \mathbf{1}) \subseteq I \times \mathcal{B}(\mathcal{H})$$

There is always the measurable projection $q : \mathbb{K} \to I$. We assume $\mathcal{H}_{I'}$ to be infinite dimensional for any $I' \subseteq I$ nonempty. Then the map $\bar{q} : \bar{\mathbb{K}} := \bigcup \mathcal{K}(\mathcal{H}_t) \otimes \mathbf{1} \setminus \{0\} \to I = (a,b)$, $\bar{q}(K_t \otimes \mathbf{1}) = t$ for $K_t \in \mathcal{K}(\mathcal{H}_t)$ is well defined.

Lemma 3.1.1 $\bar{q} : \bar{\mathbb{K}} \to I$ *is Borel measurable.*

<u>Proof:</u> For $t \in I$ take $S_t = \{s_t^i | i \in \mathbb{N}\} \subseteq \mathcal{H}_t$, $S^t \subseteq \mathcal{H}^t$ countable dense and $\mathcal{F}_t$ the family of finite dimensional subspaces of $\mathcal{H}_t$ spanned by finite subsets

of S_t having the form $\{s_t^i | i = 0, \ldots, k\}$. For $E \in \mathcal{F}_t$ let P_E be the projection onto E. For $s \in S^t$ put

$$M_{E,s,\varepsilon} := \{A \in \bar{\mathbb{K}} \mid \|[(P_E \otimes \Theta_{s,s}), A]\| < \varepsilon\}$$

Then $\bar{q}^{-1}((a,t]) = \bigcup_{t' \leq t} \mathcal{K}(\mathcal{H}_{t'}) \otimes \mathbf{1} \setminus \{0\} \subseteq \bigcap_{s \in S^t} \limsup_{E \in \mathcal{F}_t} M_{E,s,\varepsilon}$ where $\limsup_{E \in \mathcal{F}_t} M_{E,s,\varepsilon} := \bigcap_{F \in \mathcal{F}_t} \bigcup_{F \subseteq E} M_{E,s,\varepsilon}$. But also the converse inclusion holds: Suppose A is in $\mathcal{K}(\mathcal{H}_r) \otimes \mathbf{1}$ for some $r > t$ and in the r.h.s. We may assume $A = \sum_{i=1}^{k} a_i \otimes b_i \otimes \mathbf{1} \neq 0$, $a_i \in \mathcal{K}(\mathcal{H}_t)$, $b_i \in \mathcal{K}(\mathcal{H}_{[t,r)})$ all of finite rank. For each $F \in \mathcal{F}_t$, there exists $E \in \mathcal{F}_t$ s.t. $F \subseteq E$ and which contains the ranges of $a_1, \ldots, a_k$ almost. But then for any $\varepsilon > 0$ we can find $s \in S^t$ s.t. $\|[P_E \otimes \Theta_{s,s}, A]\| > \varepsilon$. Thus $A \notin M_{E,s,\varepsilon}$ and also not in $\limsup_{E \subseteq \mathcal{F}_t} M_{E,s,\varepsilon}$. So $\bar{q}^{-1}((a,t])$ is a Borel set and therefore $\bar{q}$ is Borel measurable. □

Remark 3.1.2 *A section $t \mapsto (t, K_t \otimes \mathbf{1})$ to q is Borel measurable iff $t \mapsto (K_t \otimes \mathbf{1})\xi \in \mathcal{H}$ is Borel measurable for each $\xi \in \mathcal{H}$ (by definition).*

3.1.1 L^1-Sections as Involutive Banach Algebras

Let $T = (\mathcal{H}, \mathcal{B}_0(I))$ be continuous in the sense of 2.1.6, μ a fixed continuous Borel measure on I with full support and $L^1(T)$ be the Banach space of L^1-sections to $q : \mathbb{K} \to I$. Thus $L^1(T)$ consists of families of compact operators (K_t) with $K_t \in \mathcal{K}(\mathcal{H}_t)$ s.t. $t \mapsto (K_t \otimes \mathbf{1})\xi$ is Borel measurable for each $\xi \in \mathcal{H}$ and $\int \|K_t\| d\mu < \infty$. We are going to write (K_t) for L^1-sections.

Proposition 3.1.3 *For T continuous there is a natural identification of $L^1(T)$ with $L^1(I, \mathcal{K})$ as Banach spaces.*

Proof: By Def.2.1.3 we have onb's $(e_n(t)) \subseteq \mathcal{H}_t$ and $(f_m(t)) \subseteq \mathcal{H}^t$ s.t. $t \mapsto e_n(t) \otimes f_m(t)$ are continuous maps. Define the section $t \mapsto e_{kl}(t)$ as follows: Let $\alpha_{n,m}^\xi(t) = \langle e_n(t) \otimes f_m(t), \xi \rangle$ so that $\xi = \sum_{n,m} \alpha_{n,m}^\xi(t) e_n(t) \otimes f_m(t)$. Note that $t \mapsto \alpha_{n,m}^\xi(t)$ is continuous. Put $(e_{kl}(t) \otimes \mathbf{1})\xi := \sum_m \alpha_{lm}^\xi(t) e_k(t) \otimes f_m(t)$. Then each $f_{kl} \in L^1(I, \mu)$ defines the L^1-section $t \mapsto f_{kl}(t) e_{kl}(t)$ the linear extension of which is the desired Banach space isomorphism. □

Remark 3.1.4 *(i) The dual space of $L^1(T)$ may be identified with families of trace class operators (mod $d\mu$) $(T_t)_{t \in I}$, $T_t \in \mathcal{L}^1(\mathcal{H}_t)$ s.t. $t \mapsto \|T_t\|_1 \in$*

ALGEBRAS ASSOCIATED TO CONTINUOUS T.P.

$L^\infty(I,\mu)$ (where $\|\cdot\|_1$ denotes the trace norm) and $t \mapsto tr(K_t T_t)$ is measurable for all (K_t) [1]. We call them L^∞-families.

(ii) There is a natural subspace of continuous sections in $L^1(T)$ given by the L^1-sections (K_t) s.t. $t \mapsto (K_t \otimes \mathbf{1})\xi$ is continuous for each $\xi \in \mathcal{H}$ which is the case iff f_{kl} is continuous in the above identification.

We have the following product in $L^1(T)$ (which has appearently nothing to do with the convolution in $L^1(I)$ or anything similar).

Definition 3.1.5 *The product of (K_t) and (L_s) in $L^1(T)$ is defined to be the section given by*

$$R_t = \int_a^t d\mu(s) \, [K_t(L_s \otimes \mathbf{1}) + (K_s \otimes \mathbf{1})L_t].$$

Lemma 3.1.6 *The product is well defined and $(L^1(T), \|\cdot\|_1)$ with the involution $(K_t)^* = (K_t^*)$ is a Banach-$*$-algebra.*

Proof: We only have to check submultiplicativity:

$$\begin{aligned}
\|(K_t)(L_s)\|_1 &= \int_a^b d\mu(t) \left\| \int_a^t d\mu(s) \, [K_t(L_s \otimes \mathbf{1}) + (K_s \otimes \mathbf{1})L_t] \right\| \\
&\leq \int_a^b d\mu(t) \int_a^t d\mu(s) \, (\|K_t\|\|L_s\| + \|K_s\|\|L_t\|) \\
&= \left(\int_a^b d\mu(t) \, \|K_t\| \right) \left(\int_a^b d\mu(s) \, \|L_s\| \right) \\
&= \|(K_t)\|_1 \|(L_s)\|_1.
\end{aligned}$$

□

Remark 3.1.7 (i) *Equip $L^1(I,\mu)$ with the product*

$$(f.g)(t) = \int_a^t d\mu(s)(f(s)g(t) + f(t)g(s))$$

and involution $f^*(t) := \bar{f}(t)$. Then we obtain a Banach-$*$-algebra. The map $f \mapsto (t \mapsto \int_a^t d\mu(x) \, f(x))$ is an injective $*$-homomorphism into

---

[1] of course we always mean that for a representative

$C_0(a,b]$ where the image consists of functions with derivative in $L^1(\mu)$. The L^1-norm is something between the $C^{(0)}$ and $C^{(1)}$ norm. Elements in $C_0(I) \cap L^1(I,\mu)$ are mapped to continuously differentiable functions. For $L^1(T)$ the continuous sections form a dense involutive subalgebra and may be viewed as continuously differentiable elements.

(ii) $L^1(T)^s := \{(K_t) \in L^1(T) | K_t = 0 \text{ for (almost) all } t \leq s\}$ are proper ideals for each $s \in I$. We write $(K_t)_{t \in (s,b)}$ or just $(K_t)_{(s,b)}$ for its elements.

(iii) The quotient Banach algebra $L^1(T)/L^1(T)^s$ is isomorphic to $L^1(T|(a,s))$ and we have the split exact sequence

$$0 \to L^1(T)^s \to L^1(T) \overset{\leftarrow}{\to} L^1(T|(a,s)) \to 0$$

(iv) The dependence of $L^1(T)$ on the (continuous) measure μ is less interesting for us. In case T is type I i.e. given by $\int_I^\oplus \mathcal{K}_t d\mu$, the enveloping C^*-algebra is independent of this measure and we always take μ from the direct integral in this case.

<u>Proof:</u> (i): Leibniz rule.

(ii): $dim \mathcal{H}_s = \infty$ for $s \in I$ implies that $L^1(T)^s$ is a proper subset of $L^1(T)$.

(iii): Under the correspondence in Prop.3.1.3, $L^1(T)^s$ is the complemented subspace $L^1((s,b),\mathcal{K})$. Hence $L^1(T)/L^1(T)^s \cong L^1(T|(a,s))$ on the level of Banach spaces and this isomorphy preserves the product.

(iv): One can see that using the generating cones (all of them) at the beginning of section 3.3.3.A. $\square$

Let $P_N(t) = \sum_{n=0}^N \Theta_{e_n(t), e_n(t)}$ where $\Theta_{e_n(t), e_n(t)}$ is the rank 1 projection onto $\mathbb{C}e_n(t)$ in $\mathcal{K}(\mathcal{H}_t)$. $t \mapsto P_N(t)$ is a measurable (even continuous) section.

Proposition 3.1.8 $L^1(T)$ has approximate units.

<u>Proof: Case 1 :</u> $I = (a,b)$ with $a > -\infty$. Let $(K_t) \in L^1(T)$ and define $(R_{N,t}) \in L^1(T)$ by

$$(R_{N,t}) = \tfrac{1}{\varepsilon_a}(P_N(t))_{t \in (a, a+\varepsilon)}(K_t)$$

where $\varepsilon_a = \mu([a, a+\varepsilon))$. Thus

$$R_{N,t} = \begin{cases} \frac{1}{\varepsilon_a} \int_a^t d\mu(s) \left[(P_N(s) \otimes \mathbf{1}_{[s,t)}) K_t + P_N(t)(K_s \otimes \mathbf{1}_{[s,t)}) \right] & \text{if } t \leq a+\varepsilon \\ \frac{1}{\varepsilon_a} \int_a^{a+\varepsilon} d\mu(s) \, (P_N(s) \otimes \mathbf{1}_{[s,t)}) K_t & \text{otherwise} \end{cases}$$

Because $\|(L_s)_{(a,a+\varepsilon)}\|_1 \to 0$ for $\varepsilon \to 0$ for any fixed $(L_s) \in L^1(T)$, $\|\frac{1}{\varepsilon_a}(P_N(t))_{(a,a+\varepsilon)}\|_1 = 1$ and submultiplicativity of the L^1-norm, we can neglect the first case. For the second estimate, we have for fixed s and t:

$$(P_N(s) \otimes \mathbf{1}_{[s,t)}) K_t \to K_t \text{ if } N \to \infty$$

with $t \mapsto \|K_t\|$ as majorant. By the Bochner-Lebesgue theorem (Appendix Thm.2.(ii)), we can find a zero sequence of $\varepsilon_l > 0$ and a sequence of natural numbers N_l s.t.

$$\frac{1}{\varepsilon_{l,a}}(P_{N_l}(t))_{(a,a+\varepsilon_l)}(K_t) \to (K_t)$$

Let $(x_k) \subseteq L^1(T)$ be a countable dense sequence in the unit sphere. Put $u_l := \frac{1}{\varepsilon_{l,a}}(P_{N_l}(t))_{(a,a+\varepsilon_l)}$ and choose the sequences (ε_l) and (N_l) s.t. $\|u_l x_k - x_k\| < \frac{1}{l}$ for $k = 0, \ldots, l$. Then u_l is an approximate unit.

<u>Case 2:</u> $a = -\infty$: Similar by taking $u_l := \frac{1}{\varepsilon_{l,a_l}}(P_{N_l}(t))_{(a_l, a_l+\varepsilon_l)}$ where $a_l \to -\infty$ and $\varepsilon_l \to 0$ appropriately (one could even fix $\varepsilon_l = \varepsilon$ in this case). □

Definition 3.1.9 *The C^*-algebra $\mathcal{A}(T)$ is defined to be the enveloping C^*-algebra of $L^1(T)$.*

For $T = (\mathcal{H}, \mathcal{B}_0(I))$ there is always a canonical representation π_0 of $L^1(T)$ on $\mathcal{H}$ called the regular representation defined by $\pi_0((K_t))\xi := \int_I d\mu(t) \, (K_t \otimes \mathbf{1})\xi$ where the integral is the Bochner integral in $\mathcal{H}$ [2]. The norm closure $\overline{\pi_0(L^1(T))}$ is denoted by $\mathcal{A}_r(T)$ (the reduced algebra). By determining the state space in case T of type I explicitly, we are going to see in Prop.3.2.15 that then $\mathcal{A}_r(T) = \mathcal{A}(T)$.

The reduced algebra is also equal to the concrete C^*-algebra generated by the set $\{\int_I d\mu(t) \bar{\rho}_t(K(t)) | \ t \mapsto K(t) \in L^1(I, \mathcal{K})\}$ on the Hilbert space $\mathcal{H}$

---

[2]Somewhat analogous to the regular representation λ of $C^*(E)$ which acts on a continuous Fock space $L^2(E)$ however

$$= \left(\rho(r') \int_{a'}^{r'} d\nu(s')\, \rho(s')\, u_{\tau_{r'}}[K_{\tau(r')}(L_{\tau(s')} \otimes \mathbf{1}) \right.$$
$$\left. + (K_{\tau(s')} \otimes \mathbf{1})L_{\tau(r')}]u_{\tau_{r'}}{}^*\right)$$

$$= \left(\int_{a'}^{r'} d\nu(s')\, \rho(r')\rho(s')\, [u_{\tau_{r'}}K_{\tau(r')}u_{\tau_{r'}}{}^*u_{\tau_{r'}}(L_{\tau(s')} \otimes \mathbf{1})u_{\tau_{r'}}{}^* \right.$$
$$\left. + u_{\tau_{r'}}(K_{\tau(s')} \otimes \mathbf{1})u_{\tau_{r'}}{}^*u_{\tau_{r'}}L_{\tau(r')}u_{\tau_{r'}}{}^*]\right)$$

$$= \left(\int_{a'}^{r'} d\nu(s')\, [(\rho(r')\, u_{\tau_{r'}}K_{\tau(r')}u_{\tau_{r'}}{}^*)((\rho(s')\, u_{\tau_{s'}}L_{\tau(s')}u_{\tau_{s'}}{}^* \otimes \mathbf{1})) \right.$$
$$\left. + ((\rho(s')\, u_{\tau_{s'}}K_{\tau(s')}u_{\tau_{s'}}{}^* \otimes \mathbf{1}))(\rho(r')\, u_{\tau_{r'}}L_{\tau(r')}u_{\tau_{r'}}{}^*)]\right)$$

$$= \Phi((K_t))\Phi((L_s))$$

□

Corollary 3.1.13 $\mathcal{A}(T_n(I))$ *doesn't depend on the interval I. Therefore we denote the resulting algebra by $\mathcal{A}_n$.*

3.2 The C^*-Algebra $\mathcal{A}(T)$ for T of Type I

3.2.1 Representations of $\mathcal{A}(T)$

Assume T to be given by $\int_I^\oplus \mathcal{K}_t d\mu$. Denote the rank 1 projection

$$\Theta_{exp(0|[x,y)), exp(0|[x,y))} \in \mathcal{K}(\mathcal{H}_{[x,y)})$$

$[x,y) \subseteq (a,b)$ [3] by $\epsilon_{00}(x,y)$.

Let $\mathcal{S}$ be the subspace of $L^1(T)$ introduced before Rem.3.1.10. Put for $(K_t) \in \mathcal{S}$

$$K_t^\varepsilon = \begin{cases} \frac{1}{\mu[t-\varepsilon,t)} \int_0^\varepsilon d\mu(t-x)\, K_{t-x} \otimes \epsilon_{00}(t-x,t) & \text{if } t > \varepsilon \\ 0 & \text{otherwise} \end{cases}$$

---

[3] for $(a,y) \subseteq (a,b)$ the same definition

where the integral is taken within the compact operators $\mathcal{K}(\mathcal{H}_t)$. K_t^ε is compact by Appendix Rem.3.

Lemma 3.2.1 *For each $(K_t) \in \mathcal{S}$ and $\varepsilon > 0$ we have $(K_t^\varepsilon) \in L^1(T)$ and $\int_a^b \|K_t^\varepsilon - K_t\| d\mu(t) \to 0$ if $\varepsilon \to 0$.*

<u>Proof:</u> If $f \in \int_I^\oplus \mathcal{K}_t d\mu$, then $f_{t-x} + 0|[t-x, t) \to f_t$ in norm provided $x \to 0$. Hence
$$\Theta_{exp(f_{t-x}),exp(g_{t-x})} \otimes \epsilon_{00}(t-x,t) \xrightarrow{x \to 0} \Theta_{exp(f_t),exp(g_t)}$$
It follows $\|K_t^\varepsilon - K_t\| \to 0$ for $\varepsilon \to 0$ if $K_t = \varphi(t) \Theta_{exp(f_t),exp(g_t)}$ and hence for any $(K_t) \in \mathcal{S}$. By Appendix Thm.2.(ii), the claim follows because there is a majorant. □

Remark 3.2.2 *Let $L^1(I, \mu)$ be the Banach-$*$-algebra defined in Rem.3.1.7.(i). Then $\frac{1}{\varepsilon_a}\chi_{[a,a+\varepsilon)}$ if $a > -\infty$ and $\frac{1}{\varepsilon_{a_\varepsilon}}\chi_{[a_\varepsilon,a_\varepsilon+\varepsilon)}$ are approximate units where $a_\varepsilon \to -\infty$ suitably in case of $a = -\infty$.*

<u>Proof:</u> The same as Lemma 3.1.6 and 3.1.11. □

Remark 3.2.3 *Let π be a representation of the algebraic tensor product $A \odot B$ of Banach-$*$-algebras with bounded approximate units (u_j) and (v_i). If π is a nondegenerate subcross representation on H (i.e. $\|\pi(a \otimes b)\| \leq \|a\|\|b\|$ for $a \in A$ and $b \in B$), then there are representations ϕ of A and ψ of B on the same Hilbert space H s.t. $\pi(a \otimes b) = \phi(a)\psi(b)$ defined by $\phi(a) = \lim_i \pi(a \otimes v_i)$, $\psi(b) = \lim_j \pi(u_j \otimes b)$ for $a \in A$, $b \in B$ independently of (v_i) and (u_j).*

<u>Proof:</u> The same as for C^*-algebras: For $\eta = \sum_{k=0}^n \pi(a_k \otimes b_k)\xi$ we obtain that
$$\pi(a \otimes v_i)\eta = \sum_{i=0}^n \pi(aa_k \otimes v_i b_k)\xi$$
is a Cauchy net because π is subcross and $\|v_i b_k - v_{i'} b_k\| \to 0$ for $i, i' \to \infty$. We can define $\phi(a)\eta$ and $\psi(b)\eta$ to be the respective limits. Then ϕ, ψ exist, are bounded on the dense set $\mathcal{D} = \pi(A \odot B)H$ (because (u_j) and (v_i) are bounded), independently of (v_i), (u_j) and $\pi(a \otimes b) = \phi(a)\psi(b)$ on $\mathcal{D}$, therefore also on H. □

For any representation π of $C_0(a,b]$ and $F \in C_0(a,b]$ s.t. $f = F'$ exists and lies in $L^1(dt)$ the spectral theorem may be written as $\pi(F) = \int F(x) dE(x) = \int dt\, f(t) P_t$ where $P_t = \int_t^b dE(x)$ is the spectral resolution corresponding to the spectral measure E (c.f. Appendix Def.11). This motivates the following 'spectral theorem' for $\mathcal{A}(T)$:

Proposition 3.2.4 *For each nondegenerate representation π of $L^1(T)$ on H there is a family of representations $\Pi_t : \mathcal{K}(\mathcal{H}_t) \to \mathcal{B}(H)$ s.t.*

$$\pi((K_t)) = \int_a^b d\mu(t)\, \Pi_t(K_t)$$

where the integral on the r.h.s. is taken pointwise on H in norm.

Proof: We may for simplicity assume $I = (0,1)$ (and $\mu = dt$). Consider for $t \in I$ the subalgebra $C_t \subseteq L^1(T)$ defined as

$$\{(f(x) K_t \otimes \epsilon_{00}(t, t+x))_{x \in (0, 1-t)} \mid f \in L^1(0, 1-t), K_t \in \mathcal{K}(\mathcal{H}_t)\}$$

Obviously, $C_t = \mathcal{K} \odot L^1(0, 1-t)$ where $L^1(0, 1-t)$ is considered as Banach algebra (Rem.3.1.7). The closure $\bar{C}_t$ in $L^1(T)$ is a closed subalgebra with bounded approximate unit (Rem.3.2.2). Hence π is continuous and contractive with

$$\left\| \pi\left((f(x) K_t \otimes \epsilon_{00}(t, t+x))_{x \in (0, 1-t)}\right) \right\| \leq \|K_t\| \int_0^{1-t} dx\, |f(x)|$$

which means $\pi|C_t$ being subcross and Rem.3.2.3 implies the existence of two representations

$$\Pi_t : \mathcal{K}(\mathcal{H}_t) \to \mathcal{B}(H) \quad \text{and} \quad \pi_t : L^1(0, 1-t) \to \mathcal{B}(H)$$

s.t.

$$\Pi_t(K_t) = s - \lim_{\varepsilon \to 0} \pi\left((\tfrac{1}{\varepsilon} K_t \otimes \epsilon_{00}(t, t+x))_{x \in (0, \varepsilon)}\right)$$

But for $(K_t) \in \mathcal{S}$ and $\xi \in H$ we have

$$\int_0^{1-\varepsilon} dt\ \pi\left((\tfrac{1}{\varepsilon} K_t \otimes \epsilon_{00}(t, t+x))_{x \in (0, \varepsilon)}\right) \xi =$$
$$= \pi\left(\tfrac{1}{\varepsilon} \int_0^{1-\varepsilon} dt\, (K_t \otimes \epsilon_{00}(t, t+x))_{x \in (0, \varepsilon)}\right) \xi$$
$$= \pi((K_t^\varepsilon)) \xi$$

ALGEBRAS ASSOCIATED TO CONTINUOUS T.P.

where the dt-integral is taken in the algebra $L^1(T)$. But the l.h.s. converges to $\int dt\, \Pi_t(K_t)\xi$ and the r.h.s. to $\pi((K_t))\xi$. This is the required formula for $(K_t) \in \mathcal{S}$. Because π is contractive we are done. □

Corollary 3.2.5 *The image of $(K_t) \in L^1(T)$ in $\mathcal{A}(T)$ is positive if $K_t \geq 0$ a.e. (the converse is not true).*

Corollary 3.2.6 *For $(K_t) \in L^1(T)$ and π a representation as above we have*

$$\Pi_s(K_s)\Pi_t(K_t) = \Pi_t((K_s \otimes \mathbf{1})K_t)$$

for almost all pairs $\{(s,t)|a < s < t < b\} \subseteq I^2$.

<u>Proof:</u> Up to a zero set in I^2 we have for $a < s_0 < t_0 < b$: $\Pi_{s_0}(K_{s_0})\Pi_{t_0}(K_{t_0}) = \lim_{\varepsilon \to 0} \frac{1}{\varepsilon^2} \int_{s_0}^{s_0+\varepsilon} \int_{t_0}^{t_0+\varepsilon} ds dt\, \Pi_s(K_s)\Pi_t(K_t) = \lim_{\varepsilon \to 0} \frac{1}{\varepsilon^2} \int_{s_0}^{s_0+\varepsilon} \int_{t_0}^{t_0+\varepsilon} ds dt\, \Pi_t((K_s \otimes \mathbf{1})K_t) = \Pi_{t_0}((K_{s_0} \otimes \mathbf{1})K_{t_0})$. □

3.2.2 States

Consider the dual of $L^1(T)$ given by the L^∞-families in Rem.3.1.4.(i). Because $L^1(T)$ has bounded approximate units each positive linear functional on $L^1(T)$ is bounded ([HeRo II (32.27)]). Hence it suffices to find the condition ensuring positivity for the linear functional given by an L^∞-family (T_t) of trace class operators $T_t \in \mathcal{L}^1(\mathcal{H}_t)$.

For any trace class operator $T \in \mathcal{L}^1(\mathcal{H})$ and $t \in I$ we have $\mathcal{F}^s(\int_I^\oplus \mathcal{K}_t d\mu) = \mathcal{H} = \mathcal{H}_t \otimes \mathcal{H}^t$. There is a unique trace class operator $R_t(T) \in \mathcal{L}^1(\mathcal{H}_t)$ s.t.

$$\mathrm{tr}(R_t(T)A) = \mathrm{tr}(T(A \otimes \mathbf{1}))$$

holds for all $A \in \mathcal{B}(\mathcal{H}_t)$. It is the trace class operator corresponding to the restriction of the normal linear functional to the subalgebra $\mathcal{B}(\mathcal{H}_t) \otimes \mathbf{1}$ which is again normal. If $a < s < t < b$, there is an analoguous map

$$R_s^t : \mathcal{L}^1(\mathcal{H}_t) \to \mathcal{L}^1(\mathcal{H}_s)$$

where we use $\mathcal{H}_t = \mathcal{H}_s \otimes \mathcal{H}_{[s,t)}$. R_t and R_s^t are linear, positive and trace preserving and are sometimes called relative trace maps. They provide a mean to compare T_t and T_s.

Remark 3.2.7 *(i) Let $T = T_1 \otimes T_2$, $T_1 \in \mathcal{L}^1(\mathcal{H}_t)$ and $T_2 \in \mathcal{L}^1(\mathcal{H}^t)$. Then $R_t(T) = (\text{tr } T_2) T_1$. If $\mathcal{H} \ni \xi = \sum \alpha_{nm}^\xi(t) e_n(t) \otimes f_m(t)$ and $T = \Theta_{\xi,\xi}$, then $R_t(T) = \sum_{k,n,m} \alpha_{nm}^\xi(t) \bar{\alpha}_{km}^\xi(t) \Theta_{e_n(t), e_k(t)}$.*

(ii) If $T = \Theta_{\xi,\xi}$ is of rank one, then $R_t(T)$ is of rank one iff $\xi = \xi_t \otimes \xi^t$ is a tensor and in this case $R_t(T) = \|\xi^t\|^2 \Theta_{\xi_t, \xi_t}$.

(iii) If $t \in I$, $A \in \mathcal{K}(\mathcal{H}_t)$ and $P_N \in \mathcal{K}(\mathcal{H}^t)$ is a sequence of finite dimensional projections s.t. $P_N \nearrow \mathbf{1}$, then for $T \in \mathcal{L}^1(\mathcal{H})$ we have $\text{tr}(T(A \otimes P_N)) \overset{N \to \infty}{\longrightarrow} \text{tr}(R_t(T)A)$.

Proof: It is immediate. □

Definition 3.2.8 *A family of trace class operators $T_t \in \mathcal{L}^1(\mathcal{H}_t)$ is called decreasing (constant) if*

$$R_s^t(T_t) \leq T_s \qquad (R_s^t(T_t) = T_s)$$

for $a < s < t < b$.

Now we can formulate the positivity condition for a linear functional $\psi \in L^1(T)^*$. The resulting characterization of the state space $\Omega(\mathcal{A}(T))$ is useful in the rest of this chapter.

Theorem 3.2.9 *$\psi \in L^1(T)^*$ is positive iff it is represented by a decreasing family (T_t) with $T_t \geq 0$ a.e.*

Proof: 1) Let ψ be positive ($I = (0,1)$, $\mu = dt$) and (T_t) a representing family of ψ. Cor.3.2.5 implies that $T_t \geq 0$ a.e. Let $A \in \mathcal{K}$ be a positive matrix (α_{kl}) and put

$$A_t := \sum_{kl} \alpha_{kl} e_{kl}(t)$$

($e_{kl}(t)$ as in the proof of Prop.3.1.3.) where we mean the norm limit of the sequence $\sum_{k,l \leq n} \alpha_{kl} e_{kl}(t)$. (Note, however, that if $R = (r_{ij})$ and $T = (t_{ij})$ are Hilbert-Schmidt, then any submatrices are again Hilbert-Schmidt with smaller HS-norm. Thus the series $\sum_{ikl} r_{ik} \alpha_{kl} t_{li}$ is absolutely convergent and hence the above sum is weakly (unconditionally) subseries convergent. The

Orlitz-Pettis theorem [Die 84 p.24] implies that the same holds in norm so the order in the sum doesn't matter).

It is clear that $t \mapsto A_t$ is in $L^1(T)$. We define for $0 < s_0 < t_0 < 1$ and $\varepsilon > 0$ sufficiently small

$$a_\varepsilon := (\tfrac{1}{\varepsilon} A_s)_{s \in (s_0, s_0 + \varepsilon)}$$

By Cor.3.2.5, $\psi(a_\varepsilon) \geq 0$ in $\mathcal{A}(T)$.

Next we find, for $0 < s < t < 1$, by iterating the proof of Prop.2.1.20, onb's $(e_n(s))$, $(f_m(s,t))$, $(g_p(t))$ of $\mathcal{H}_s$, $\mathcal{H}_{[s,t)}$, $\mathcal{H}^t$ s.t. $e_n(s) \otimes f_m(s,t) \otimes g_p(t)$ is continuous for arbitrary $n, m, p \in \mathbb{N}$ in s and t, $s < t$. Let $P_{N,[s,t)} = \sum_{m=0}^{N} \Theta_{f_m(s,t), f_m(s,t)}$ and define for $\varepsilon < t_0 - s_0$, $\varepsilon' < 1 - t_0$

$$a_{\varepsilon, \varepsilon', N} := (\tfrac{1}{\varepsilon\varepsilon'} \int_{s_0}^{s_0+\varepsilon} ds\, A_s \otimes P_{N,[s,t)})_{t \in (t_0, t_0+\varepsilon')}$$

Then $\pi(a_\varepsilon) - \pi(a_{\varepsilon, \varepsilon', N}) \geq 0$ for any representation (By Cor.3.2.6 $\Pi_t(A_s \otimes P_{N,[s,t)}) \leq \Pi_s(A_s)$ holds for almost all $\{(s,t) | s \in [s_0, s_0+\varepsilon], t \in [t_0, t_0+\varepsilon']\}$. Hence $\pi(a_{\varepsilon, \varepsilon', N}) \leq \pi(a_\varepsilon)$ by integration).

Now $s \mapsto \mathrm{tr}(T_s A_s)$ is in $L^1(I)$. Therefore (Appendix Rem.6)

$$(*) \quad \psi(a_\varepsilon) = \tfrac{1}{\varepsilon} \int_{s_0}^{s_0+\varepsilon} ds\, \mathrm{tr}(T_s A_s) \xrightarrow{\varepsilon \to 0} \mathrm{tr}(T_{s_0} A_{s_0})$$

for almost all $s_0 \in I$. Taking $\mathcal{D}_+ = \{A^1, A^2, \ldots\} \subseteq \mathcal{K}_+$ a countable dense subset, we may change the representative (T_t) on a zero set in order that $(*)$ holds for each $A \in \mathcal{D}_+$ and any $s_0 \in I$. Because $A \mapsto a_\varepsilon$ is isometric for each ε it then holds for any compact positive A. In particular, T_{s_0} is positive for each $s_0 \in I$ (not only a.e.). Furthermore

$$\begin{aligned}
\psi(a_{\varepsilon,\varepsilon',N}) &= \tfrac{1}{\varepsilon\varepsilon'} \int_{s_0}^{s_0+\varepsilon} ds \int_{t_0}^{t_0+\varepsilon'} dt\, \mathrm{tr}(T_t(A_s \otimes P_{N,[s,t)})) \\
&= \tfrac{1}{\varepsilon\varepsilon'} \int_{t_0}^{t_0+\varepsilon'} dt\, \mathrm{tr}\left(T_t\left(\int_{s_0}^{s_0+\varepsilon} ds\, A_s \otimes P_{N,[s,t)}\right)\right) \\
&\xrightarrow{\varepsilon' \to 0} \tfrac{1}{\varepsilon} \mathrm{tr}\left(T_{t_0} \underbrace{\left(\int_{s_0}^{s_0+\varepsilon} ds\, A_s \otimes P_{N,[s,t)}\right)}_{\geq 0}\right)
\end{aligned}$$

where the indicated integral lives in $\mathcal{K}(\mathcal{H}_t)$ and is positive. If $\varepsilon \to 0$, this converges to
$$\mathrm{tr}(T_{t_0}(A_{s_0} \otimes P_{N,[s_0,t_0)}))$$
for all s_0 using the continuity in s and for $N \to \infty$ by Rem.3.2.7.(iii) to
$$\mathrm{tr}(R_{s_0}^{t_0}(T_{t_0})A_{s_0})$$
Starting with N, we choose ε_N and then ε'_N in order to obtain zero sequences s.t.
$$\psi(a_{\varepsilon_N} - a_{\varepsilon_N,\varepsilon'_N,N}) \stackrel{N \to \infty}{\longrightarrow} \mathrm{tr}((T_{s_0} - R_{s_0}^{t_0}(T_{t_0}))A_{s_0}) \geq 0$$
for any positive $A \in \mathcal{K}$.

2) Suppose (T_t) is a decreasing family corresponding to ψ. Then the family defined by $T_t^m := R_t^{k/2^m}(T_{k/2^m})$, k the unique natural number s.t. $t \in ((k-1)/2^m, k/2^m]$, defines a positive linear functional ψ_m (it is a sum of positive normal linear functionals in the representations $\pi_{k/2^m}$ defined before Rem.3.3.5) s.t. $\psi_m(y) \stackrel{m \to \infty}{\longrightarrow} \psi(y)$ for each continuous $y \in L^1(T)$ in the sense of Rem.3.1.4.(ii). Such y form a dense subset so $\psi \geq 0$. □

Corollary 3.2.10 *Let φ be a positive linear functional on $\mathcal{A}(T)$ and (T_t) as in Thm.3.2.9. Then*

(i) $\|\varphi\| = \lim_{t \to a} \|T_t\|_1$ where $\|\cdot\|_1$ is the trace norm.

(ii) There exists a unique representative (T_t) of φ s.t. $t \mapsto \|T_t\|$ is right continuous.

<u>Proof:</u> (i): $\varphi(u_l) = \frac{1}{\varepsilon_{l,a}} \int_a^{a+\varepsilon_l} d\mu(t)\, \mathrm{tr}(T_t P_{N_l}(t))$ for the approximate unit in the proof of Prop.3.1.8. This converges to $\lim_{t \to 0} \|T_t\|_1$ if $N_l \nearrow \infty$ sufficiently fast.

(ii) Let (T_t) as above. Then $t \mapsto \|T_t\|$ is monotone decreasing and can only have countably many jumps. Let t_0 be such a point and consider the ideal $\mathcal{A}^{t_0}(T) \cong \mathcal{K} \otimes \mathcal{A}(T|(t_0,b)) = \overline{L^1(T)^{t_0}}$. For an approximate unit (u_l) in $\mathcal{A}(T|(t_0,b))$ like in Prop.3.1.8, the map $\bar{\varphi}(K) = \lim \varphi(K \otimes u_l)$ is a positive linear functional on $\mathcal{K}(\mathcal{H}_{t_0})$ represented by a trace class operator T'_{t_0}. Redefining at most countably often, we obtain the desired representative. □

Proposition 3.2.11 *Let A be a C^*-algebra, $I \triangleleft A$ a closed ideal and φ a positive linear functional on A. Then there is a unique decomposition $\varphi = \varphi_1 \oplus \varphi_2$ s.t. $\|\varphi_1\| = \|\varphi|I\|$ and $\varphi_2(I) = 0$.*

Proof: [Dix 77, 2.11.7] □

Let $\mathcal{A}^s$ be the closure of $L^1(T)^s$ in $\mathcal{A}(T)$ where $s \in I$. It is a closed ideal.

Corollary 3.2.12 *Let φ be a positive linear functional on $\mathcal{A}(T)$, (T_t) as in Cor.3.2.10.(ii) and $s \in I$ fixed. Define positive linear functionals φ_1, φ_2 by*

$$T_t^1 = \begin{cases} R_t^s(T_s) & \text{if } t < s \\ T_t & \text{otherwise} \end{cases}$$

and

$$T_t^2 = \begin{cases} T_t - R_t^s(T_s) & \text{if } t < s \\ 0 & \text{otherwise} \end{cases}$$

Then $\varphi = \varphi_1 + \varphi_2$ is the unique decomposition as in the above proposition for $A = \mathcal{A}(T)$ and $I = \mathcal{A}^s$.

Proof: $\|\varphi_1\| = \|\varphi|\mathcal{A}^s\|$ by Cor.3.2.10 and $\varphi_2(\mathcal{A}^s) = 0$ is obvious. □

Corollary 3.2.13 *(i) Let φ be a pure state on $\mathcal{A}(T)$ and (T_t) as in Cor.3.2.10.(ii). Then there is a unique $r \in (a, b]$ s.t. $T_s = R_s^t(T_t)$ for any $a < s < t < r$ and $T_t = 0$ for $t \geq r$.*

(ii) For $\varphi, \psi \in \mathcal{A}(T)_+^*$ represented by (T_t) and (S_t) we have $\varphi \geq \psi$ if $T_t \geq S_t$ for each $t \in I$ and $T_s - R_s^t(T_t) \geq S_s - R_s^t(S_t)$ for $s < t$.

Proof: (i): Otherwise we could decompose φ according to the preceding Corollary into a nontrivial convex combination.

(ii): ($I = (0,1)$) Let $\psi_m := \sum_i \psi_m^i$ where the ψ_m^i are positive linear functionals given by

$$T_t^{i,m} = \begin{cases} R_t^{i/2^m}(T_{i/2^m}) - R_t^{i+1/2^m}(T_{i+1/2^m}) & \text{if } t \leq \frac{i}{2^m},\ i \leq 2^m \\ 0 & \text{otherwise} \end{cases}$$

(i.e. as in the proof of Thm.3.2.9 part 2)) and φ_m analogously. The assumption implies that $\varphi_m^i \geq \psi_m^i$ for each i and m. But $\varphi_m(x) \to \varphi(x)$ and $\psi_m(x) \to \psi(x)$ for each $x \in \mathcal{S}$ thus for each $x \in \mathcal{A}(T)$. □

Remark 3.2.14 *It is not true that each constant family (T_t) is a normal state w.r.t. the regular representation π_0 (otherwise $\mathcal{A}(T)$ would be type I).*

Proof: Let $(t_k) \subseteq (0,1) = I$ be a strictly increasing sequence $t_k \nearrow 1$. We have $\mathcal{H} = \bigotimes_{k=0}^{\infty} (\mathcal{H}_{[t_k, t_{k+1})}, \Omega_k)$ with $\Omega_k := exp(0|[t_k, t_{k+1}))$. We take a sequence $\xi_k \in \mathcal{H}_{[t_k, t_{k+1})}$, $\|\xi_k\| = 1$ s.t. $\sum_k |\langle \xi_k, \Omega_k \rangle - 1|$ is divergent. Specifically, choosing onb's $(f_m(t_k, t_{k+1}))$ in $\mathcal{H}_{[t_k, t_{k+1})}$, consider $\xi_{(m_1, \ldots, m_p)} = e_n(t_1) \otimes f_{m_1}(t_1, t_2) \otimes \ldots \otimes f_{m_p}(t_p, t_{p+1}) \otimes exp(0|[t_{p+1}, 1))$. Then $\xi_{(m_1, \ldots, m_p)} \stackrel{w}{\to} 0$ if at least one of the indices tends to infinity. Let $S = \{s_n | n \in \mathbb{N}\}$ be a countable dense subset of the unit sphere of $\mathcal{H}$. Then we can choose a sequence $m_1, m_2, \ldots$ inductively s.t. for $M_p := (m_1, \ldots, m_p)$ we have $|\langle \xi_{M_p}, s_i \rangle| \leq \frac{1}{p}$ for all $i \leq p$. Therefore $\xi_{M_p} \stackrel{w}{\to} 0$ if $p \to \infty$. We can define (T_t) constant by putting $T_t = R_t^{t_p}(\Theta_{\xi_{M_p}, \xi_{M_p}})$ provided $t_p > t$. The resulting family is in fact independent of the choice of $t_p > t$. The corresponding state ω is the pointwise limit of the vector states $\omega_p = \langle \xi_{M_p}, \cdot \xi_{M_p} \rangle$. Suppose ω is normal, then it must be a vector state since it is pure on $\mathcal{A}(T|(0, t_k))$ for each $k \in \mathbb{N}$ hence pure itself (using Prop.3.2.15). But this is not true by construction. □

3.2.3 Ideals and Exact Sequences

We continue to write $\mathcal{A}^s$ for the closure of $L^1(T)^s$ in $\mathcal{A}$.

Proposition 3.2.15 *The regular representation π_0 of $\mathcal{A}(T)$ is faithful and irreducible.*

Proof: ($I = (0,1)$) 1): π_0 is faithful: Let φ be a state on $\mathcal{A}(T)$ and (T_t) as in Cor.3.2.10.(ii). Then we have to show that (∗) $|\varphi(x)| \leq \|\pi_0(x)\|$ holds for x in a dense subset of $\mathcal{A}(T)$. Without loss assume φ to be pure. By Cor.3.2.13 there is a unique $r \in I \cup \{1\}$ s.t. $T_s = R_s^t(T_t)$ for $0 < s < t < r$ and $T_t = 0$ for $t \geq r$. Therefore if $\mathcal{A}(0, r)$ is the closure of $L^1(T|(0, r))$, then $\varphi|\mathcal{A}(0, r)$ is pure and constant. Let $P_l \nearrow \mathbf{1}$ be an increasing sequence of finite dimensional

ALGEBRAS ASSOCIATED TO CONTINUOUS T.P.   49

projections in $\mathcal{H}^t$ then $\|\pi_0((K_t))\| \geq \lim_{l \to \infty} \|(\mathbf{1} \otimes (\mathbf{1} - P_l))\pi_0((K_t))\| = \|\pi_0((K_t)_{(0,r)})\|$. It is therefore enough to check $(*)$ for $x \in L^1(T|(0,r))$. But if $x = (K_t)_{(0,r)}$, then we have $\lim_{s \nearrow r}(K_t)_{(0,s)} = x$. For each $\varepsilon > 0$ there is a normal state φ_ε defined by $T_t^\varepsilon = R_t(T_{r-\varepsilon} \otimes \Theta_{\xi^{r-\varepsilon},\xi^{r-\varepsilon}})$, $\xi^{r-\varepsilon} \in \mathcal{H}^{r-\varepsilon}$ any unit vector. φ_ε has the property $\varphi_\varepsilon((K_t)_{(0,r-\varepsilon)})) = \varphi((K_t)_{(0,r-\varepsilon)})$. Hence $\varphi_\varepsilon((K_t)_{(0,r)}) = \varphi((K_t)_{(0,r)}) + o(\varepsilon)$ and $|\varphi_\varepsilon(x)| \leq \|\pi_0(x)\|$ using normality of φ_ε. Thus $|\varphi(x)| \leq \|\pi_0(x)\|$.

2): π_0 is irreducible: For $f,g,h,k \in \int_I^\oplus \mathcal{K}_t d\mu$ ($d\mu = dt$) we have

$$\langle exp(h), \pi_0\left((\tfrac{1}{\varepsilon}\Theta_{exp(f_t),exp(g_t)})_{(b-\varepsilon,b)})\right) exp(k)\rangle = \tfrac{1}{\varepsilon}\int_{1-\varepsilon}^1 dt\; e^{\langle h_t,f_t\rangle+\langle g_t,k_t\rangle+\langle h^t,k^t\rangle}$$

which converges to $\langle exp(h), (\Theta_{exp(f),exp(g)})exp(k)\rangle$ for $\varepsilon \to 0$. Thus $\overline{\pi_0(\mathcal{A}(T))}^w$ contains the compacts. □

If we write $\int d\mu(t)\; K_t \otimes \mathbf{1}$ for $(K_t) \in L^1(T)$ (sometimes $\int d\mu(t)\; K_t$) in the sequel, we usually mean the pointwise integral in the regular or any other faithful representation.

Proposition 3.2.16 *Each constant family (T_t) with $\|T_t\|_1 = 1$ and $T_t \geq 0$ induces a faithful representation of $\mathcal{A}(T)$. The corresponding state φ restricted to $\mathcal{A}(a,s)$ is normal w.r.t. this subalgebra for each $s < b$.*

Proof: Clear. □

Proposition 3.2.17 *For each $s \in I$ there is a split exact sequence*

$$0 \to \mathcal{A}^s \to \mathcal{A}(T) \overset{\leftarrow}{\to} \mathcal{A}(a,s) \to 0$$

In particular, for each $s \in I$:

$$\mathcal{A} \cong \mathcal{A}^s \oplus \mathcal{A}(a,s) = \mathcal{K} \otimes \mathcal{A}(s,b) \oplus \mathcal{A}(a,s) \otimes \mathbf{1}$$

Proof: Consider the closure of the split exact sequence

$$0 \to L^1(T)^s \to L^1(T) \overset{\leftarrow}{\to} L^1(T|(a,s)) \to 0$$

in the regular representation π_0 of $L^1(T)$. Let π_0' be the regular representation of $L^1(T|(s,b))$ and π_0'' be the regular representation of $L^1(T|(a,s))$.

Then we have $\overline{\pi_0(L^1(T)^s)} \cong_u \mathcal{K} \otimes \overline{\pi'_0(L^1(T|(s,b)))}$ and $\overline{\pi_0(L^1(T|(a,s)))} \cong_u \overline{\pi''_0(L^1(T|(a,s)))} \otimes \mathbf{1}$. Thus $0 \to L^1(T)^s \to L^1(T)$ extends to an inclusion $0 \to \mathcal{A}^s \stackrel{i}{\hookrightarrow} \mathcal{A}(T)$ and $L^1(T|(a,s)) \to L^1(T)$ extends to an inclusion $\mathcal{A}(a,s) \stackrel{k}{\hookrightarrow} \mathcal{A}(T)$. $L^1(T) \to L^1(T|(a,s)) \to 0$ extends to $\mathcal{A}(T) \to \mathcal{A}(a,s) \to 0$ because each representation of $L^1(T|(a,s))$ gives one of $L^1(T)$ by composition. For exactness in the middle, suppose $\overline{\pi_0(L^1(T)^s)} \cap \overline{\pi_0(L^1(T|(a,s)))} \neq \{0\}$. Let $\mathcal{D} := \{x \in L^1(T|(a,s)) \mid \exists s' < s : x = (K_t)_{(a,s')}\}$. Then $\mathcal{D}$ is dense in $L^1(T|(a,s))$ and there are sequences $(a_k) \subseteq L^1(T)^s$, $(b_k) \subseteq \mathcal{D}$ s.t. $\|a_k\| = \|b_k\| = 1$ but $\|a_k - b_k\| \to 0$. Consider $K \in \mathcal{K}(H \otimes H)$, H separable infinite dimensional. Let $P_l \nearrow \mathbf{1}$ be an increasing sequence of finite dimensional projections. Then $K(\mathbf{1} \otimes P_l) \stackrel{l \to \infty}{\longrightarrow} K$ and if $R \in \mathcal{K}$, then $\|K - R \otimes \mathbf{1}\| \geq \|K(\mathbf{1} \otimes (\mathbf{1} - P_l)) - (R \otimes \mathbf{1})(\mathbf{1} \otimes (\mathbf{1} - P_l))\| \stackrel{l \to \infty}{\longrightarrow} \|R\|$. It follows that $\|a_k - b_k\| \geq \|b_k\| = 1$ which is a contradiction. □

Lemma 3.2.18 π_0 is faithful as a representation of $L^1(T)$.

Proof: Suppose $\pi_0((K_t)) = 0$ for some nonzero (K_t). Using the same sequence of projections P_l in $\mathcal{H}^s$ as above, we obtain $\pi_0((K_t)_{(a,s)}) = 0$ for each $s \in I$. But for any fixed $\xi \in \mathcal{H}$ Appendix Rem.6 implies $\lim_{\varepsilon \to 0} \frac{1}{\mu([s,s+\varepsilon))} \pi_0((K_t)_{(s,s+\varepsilon)}) \xi = (K_s \otimes \mathbf{1}) \xi$ for almost all $s \in I$ which is impossible. □

Corollary 3.2.19 (i) Each proper ideal in $\mathcal{A}(T)$ is of the form $\mathcal{A}^s$ for some $s \in I$.

(ii) $\mathcal{A}(T)$ is not type I.

Proof: (i): It suffices to show this for primary ideals (i.e. kernels of irreducible representations). Let φ be a pure state with (T_t) and $r \in I$ as in Cor.3.2.13.(i). Then φ induces a faithful representation of $\mathcal{A}(a,r)$ by Prop.3.2.16. The kernel contains $\mathcal{A}^r$. Using the topological direct sum decomposition $\mathcal{A} = \mathcal{A}(a,r) \oplus \mathcal{A}^r$ just established, the kernel is exactly $\mathcal{A}^r$.

(ii): Suppose $\mathcal{A}(T)$ is type I. Because π_0 is irreducible [Ped 79 6.1.5] implies that $\pi_0(\mathcal{A}(T)) = \overline{\pi_0(L^1(T))}$ contains the compacts. Let $\mathcal{D} := \{x \in L^1(T) \mid \exists s < b : x = (K_t)_{(a,s)}\} \subseteq L^1(T)$. For any $K \in \mathcal{K}(\mathcal{H})$ there is a sequence $(b_k) \subseteq \mathcal{D}$ s.t. $b_k \to K$. But $\|K - b_k\| \geq \|b_k\|$ by the same argument as in 3.2.17. Thus $K = 0$. □

3.3 Automorphisms and Endomorphisms

3.3.1 Ideal Preserving Automorphisms

Let $I = (a, b)$, $I' = (a', b')$ if two algebras $\mathcal{A}(T)$ and $\mathcal{A}(T')$ are involved. Note that the ideal space of $\mathcal{A}(T)$ may be identified with $\bar{I} = [a, b]$, $\mathcal{A}^a := \mathcal{A}(T)$, $\mathcal{A}^b := 0$.

Remark 3.3.1 *Each isomorphism $\varphi : \mathcal{A}(T) \to \mathcal{A}(T')$ induces a unique continuous invertible function $\sigma : [a, b] \to [a', b']$ s.t. $\sigma(a) = a'$.*

Proof: $\varphi(\mathcal{A}^s)$ must be an ideal in $\mathcal{A}(T')$ hence of the form $\mathcal{A}^{\sigma(s)}$. σ defined as such is monotone and invertible, therefore continuous. □

In order to study a class of isomorphisms between $\mathcal{A}(T)$'s the following set of states is useful.

Definition 3.3.2 *Denote by $\check{S} = \check{S}(T)$ the following set of states on $\mathcal{A}(T)$:*

$$\check{S} := \{\omega \in \Omega(\mathcal{A}(T)) \mid \omega|\mathcal{A}(a, s) \text{ is pure and } \|\omega|\mathcal{A}^s\| = 1 \text{ for each } s \in I\}$$

(Elements of $\check{S}$ may be called locally pure states).

There is a correspondence between factorizable unit vectors and locally pure states.

Proposition 3.3.3 *Let $\omega \in \Omega(\mathcal{A}(T))$ be a state and (T_t) as in Cor.3.2.10.(ii). Then $\omega \in \check{S}$ iff $\omega|\mathcal{A}(a, t)$ is a pure and normal state w.r.t. the regular representation of $\mathcal{A}(a, t)$ for each $t \in I$. There exists an a.e. unique strongly $\mathcal{S}$-measurable $f \in \prod_{s \in (a,t)} \mathcal{K}_s$ s.t. $f_t \in \int_{(a,t)}^{\oplus} \mathcal{K}_s d\mu(s)$ for each $t \in (a, b)$ and $T_t = e^{-\|f_t\|^2} \Theta_{exp(f_t), exp(f_t)}$.*

Proof: Let $\omega \in \check{S}$ and (T_t) as in Cor.3.2.10.(ii). Then we have for fixed $t \in I$ and $s \leq t$: $T_s = R_s^t(T_t)$. Thus $(T_s)_{s \in (a,t)}$ defines a normal state on $\mathcal{A}(a, t)$ w.r.t. its regular representation given by the density matrix T_t. Because the regular representation is irreducible (Prop.3.2.15) (T_t) must be a vector state $T_t = \Theta_{\xi_t, \xi_t}$. By Rem.3.2.7.(ii), ξ_t is factorizable in the sense that for $a < t_1 < \ldots < t_p < t < b$ we can find unit vectors $\xi_{t_1} \in \mathcal{H}_{t_1}, \xi_{[t_1, t_2)} \in$

$\mathcal{H}_{[t_1,t_2)}, \dots, \xi_{[t_p,t)} \in \mathcal{H}_{[t_p,t)}$ (unique up to a scalar of absolute value 1) s.t. $\xi_t = \xi_{t_1} \otimes \xi_{[t_1,t_2)} \otimes \dots \otimes \xi_{[t_p,t)}$. Now Thm.2.2.4 implies the existence of the family $f_t \in \int_{(a,s)}^{\oplus} \mathcal{K}_s d\mu$ by the same argument as in Lemma 2.1.23 with the asserted properties.

If conversely $\omega|\mathcal{A}(a,t)$ is a pure and normal state for each $t \in I$, then (T_t) must be constant. $\square$

Definition 3.3.4 *An isomorphism $\phi : \mathcal{A}(T) \to \mathcal{A}(T')$ is called diagonal if $\phi(\mathcal{A}(T|(a,t))) = \mathcal{A}(T'|(a',\sigma(t)))$ for each $t \in I$.*

We observe that any diagonal isomorphism ϕ induces a bijection $\phi^* : \check{S}(T') \to \check{S}(T)$. Moreover, ϕ^* is distance preserving. This fact can be used to show that quite often there exist no diagonal isomorphisms between different algebras of type $\mathcal{A}(T)$.

The above notion may be used to determine certain canonical ideal preserving automorphism of $\mathcal{A}(T)$. First remark that the unitary group in the multipliers of a nonunital C^*-algebra splits into the product of the central unitaries and the ones acting as nontrivial inner automorphisms. Each central multiplier becomes a multiplication with a complex number in each irreducible representation thus may be viewed as a function on the prime spectrum of the algebra. So each central multiplier of $\mathcal{A}(T)$ is given by a bounded complex valued function φ on $(a,b]$ and acts by diagonal multiplication m_φ with φ on the representation space of

$$\bar{\pi} : \mathcal{A}(T) \to \mathcal{B}\left(\int_I^\oplus \mathcal{H}_t d\mu(t)\right) \qquad \bar{\pi} = \int^\oplus \pi_t \, d\mu(t)$$

where $\pi_t : \mathcal{A}(T) \to \mathcal{B}(\mathcal{H}_t)$ is defined by

$$\pi_t((K_t))\xi := \int_a^t d\mu(s) \, (K_s \otimes \mathbf{1}_{[s,t)})\xi$$

In the irreducible representation π_t, m_φ becomes multiplication with $\varphi(t)$.

Remark 3.3.5 *There are no nontrivial central multipliers of $\mathcal{A}$.*

Proof: Suppose $\varphi : (a,b] \to \mathbb{C}$ belongs to a central multiplier. For $a < s < t < b$ let $x \in \mathcal{A}(a,s)$. Then x is multiplied by $\varphi(t)$ in the representation π_t and by $\varphi(s)$ in the representation π_s. $\square$

ALGEBRAS ASSOCIATED TO CONTINUOUS T.P. 53

Theorem 3.3.6 *(i) $\alpha \in Aut(\mathcal{A}(T))$ is diagonal ideal preserving iff there exists D in the unitary diagonal operators of $\int_I^\oplus \mathcal{K}_t d\mu$ and $f \in \prod \mathcal{K}_s$ measurable s.t. $f_t \in \int_{(a,t)}^\oplus \mathcal{K}_t d\mu$ with the property*

$$\alpha(x) = W(D_t, f_t) \, x \, W(D_t, f_t)^* \quad for \; x \in \mathcal{A}(a,t)$$

for each $a < t < b$ and understood in the representation π_t, $W(A, f) = W(A, f, 1)$ the Weyl unitary considered in Lemma 2.1.15.

(ii) $\alpha \in Aut(\mathcal{A}(T))$ is diagonal and universally weakly inner iff there exists a diagonal unitary D, and $f \in \int_I^\oplus \mathcal{K}_t d\mu$ s.t. $U = W(D_t, f_t)$ in the representation π_t.

Proof: (i): We only have to show one direction. Let Ω_c be the set of constant states (i.e. the associated families (T_t) as in 3.2.10.(ii) are constant) and α diagonal as well as ideal preserving. For $\omega, \omega' \in \Omega_c$, put $\omega \cdot \omega' = 1 - \frac{1}{4}\|\omega - \omega'\|^2$, $\omega_t \cdot \omega'_t = 1 - \frac{1}{4}\|\omega_t - \omega'_t\|^2$ where $\omega_t := \omega|\mathcal{A}(a,t)$. Because α is diagonal and ideal preserving it induces isomorphisms $\alpha(t) : \mathcal{A}(a,t) \to \mathcal{A}(a,t)$ s.t. $\alpha(t)_*(\omega_t) = (\alpha_*\omega)_t$ and, on the dual of the C^*-algebra $\mathcal{A}(T)$, this map is an isometry, preserving Ω_c and $\check{S}$. Because $\|\alpha(t)_*\omega_t - \alpha(t)_*\omega'_t\| = \|\omega_t - \omega'_t\|$ we obtain $(\alpha(t)_*\omega_t) \cdot (\alpha(t)_*\omega'_t) = \omega_t \cdot \omega'_t = |\langle T_t^{1/2}, T_t'^{1/2}\rangle|^2$ ($\langle \cdot, \cdot \rangle$ the Hilbert-Schmidt scalar product) provided T_t and T'_t are of rank 1. If T_t and T'_t are of rank 1, then so are their images under $\alpha(t)_*$. Thus for each $t \in I$ we get a projective isomorphism $\mathbf{T}_t : P\mathcal{H}_t \to P\mathcal{H}_t$ s.t. $|\langle \mathbf{T}_t[\xi_t], \mathbf{T}_t[\eta_t]\rangle| = |\langle[\xi_t], [\eta_t]\rangle|$. Because α preserves $\check{S}$ it follows from Prop.3.3.3 that for $exp(f) \in \mathcal{H}_t$ we have $\mathbf{T}_t[exp(f)] = [exp(f')]$ for some $exp(f') \in \mathcal{H}_t$. By Wigners Theorem [Par 92 sec.14], $\mathbf{T}_t$ has a unitary or antiunitary lift U_t and therefore U_t preserves exp-vectors up to a phase. If U_t is unitary, then $U_t = W(D_t, f_t, c_t)$ for some $f_t \in \int_{(a,t)}^\oplus \mathcal{K}_s \, d\mu(s)$ and a unitary D_t on that space. If U_t is antiunitary, then $U_t = J_t W(D_t, f_t, c_t)$ where J_t is a conjugation preserving exp-vectors (c.f. Rem.2.2.6). Because α is linear only the first case occurs.

Now suppose $a < s < t < b$ and let $\xi_t \in \mathcal{H}_t$, $\xi^t \in \mathcal{H}^t$, $\xi_s \in \mathcal{H}_s$ and $\xi_{[s,t)} \in \mathcal{H}_{[s,t)}$ be all unit vectors s.t. $\xi_t = \xi_s \otimes \xi_{[s,t)}$. Then $\xi_t \otimes \xi^t$ defines a pure state on $\mathcal{A}(T)$ which is pure on $\mathcal{A}(a,t)$ and $\mathcal{A}(a,s)$. By assumption this property is preserved under α. Hence there are $\xi'_s \in \mathcal{H}_s$ and $\xi'_{[s,t)} \in \mathcal{H}_{[s,t)}$ s.t. $\mathbf{T}_t[\xi_t] = [\xi'_s \otimes \xi'_{[s,t)}]$. We may define $T_{[s,t)}[\xi_{[s,t)}] = [\xi'_{[s,t)}]$ and have $T_t =$

$T_s \otimes T_{[s,t)} = T_{s_1} \otimes T_{[s_1,s_2)} \otimes \ldots \otimes T_{[s_k,t)}$ for any partition $a < s_1 < \ldots < s_k < t$. But $T_t = [W(D_t, f_t)]$ and it follows from the linear independence of expvectors as in the proof of 3.3.11 that D_t maps $\int_{[s,t)}^\oplus \mathcal{K}_r \, d\mu(r)$ into itself for any subinterval which means that D_t is diagonal. D_t, f_t extend D_s, f_s for $s < t$ (as in Lemma 2.1.23).

(ii): Let α be diagonal and given by $Ad(U)$ in each faithful representation. Using the central open projection belonging to an ideal it follows that α must be ideal preserving. Taking the regular representation, it is clear that the families D_t and f_t in (i) must fit together to a unitary $W(D, f)$ with $f \in \mathcal{H}$. The converse is obvious. □

Corollary 3.3.7 *There are no nontrivial diagonal and at the same time inner automorphisms of $\mathcal{A}(T)$.*

Proof: No nontrivial factorizable unitary is a multiplier. □

We mention that the nuclearity of $\mathcal{A}_n = \mathcal{A}(T_n)$ follows from the nuclearity of $C^*(E_n)$, the crossed product representation in sec.4.3 and Takai duality. There is also a direct proof which covers the case $\mathcal{A}(T)$ as well.

Consider for $k \in \mathbb{N}$ the Hilbert space $\mathbb{C}^k \otimes \mathcal{H}$ as a tensor decomposition T^k over $\mathcal{B}_0([a,b))$ with the 'atom' $\mathbb{C}^k$ at a. We have $\mathcal{A}(T^k) = M_k(\mathcal{A}(T))$ and the positive linear functionals of $\mathcal{A}(T^k)$ are given by decreasing families of trace class operators (T_t), $T_t \in \mathcal{L}^1(\mathbb{C}^k \otimes \mathcal{H}_t)$, as in Cor.3.2.10.(ii). Let $\underline{t}$ denote the finite sequence $a = t_0 < t_1 \ldots < t_n < b$. Define contractions $\varphi_{\underline{t}} : \mathcal{A}(T)^* \to \mathcal{A}(T)^*$ as acting on families of trace class operators by

$$\varphi_{\underline{t}}((T_t)) = \begin{cases} R_t^{t_i}(T_{t_i}) & \text{if } t \in [t_{i-1}, t_i), \ i = 1, \ldots, n \\ 0 & \text{otherwise} \end{cases}$$

and $\varphi_{\underline{t}}^{(k)} : M_k(\mathcal{A}(T))^* \to M_k(\mathcal{A}(T))^*$ by the same formula applied to families (T_t), $T_t \in \mathcal{L}^1(\mathbb{C}^k \otimes \mathcal{H}_t)$.

Lemma 3.3.8 $\varphi_{\underline{t}}^{(k)} = id_{M_k} \otimes \varphi_{\underline{t}}$, *in particular $\varphi_{\underline{t}}$ is a completely positive contraction.*

ALGEBRAS ASSOCIATED TO CONTINUOUS T.P.

<u>Proof:</u> Let K, L be Hilbert spaces. Consider the relative trace maps $R_1 : \mathcal{L}^1(K \otimes L) \to \mathcal{L}^1(K)$ and $R_2 : \mathcal{L}^1(\mathbb{C}^k \otimes K \otimes L) \to \mathcal{L}^1(\mathbb{C}^k \otimes K)$. Then

$$R_2\Big(\sum_{ijklmn} \alpha_{ijk,lmn}\Theta_{e_i\otimes f_j\otimes g_k, e_l\otimes f_m\otimes g_n}\Big) = \sum_{ijlm}\Big(\sum_k \alpha_{ijk,lmk}\Theta_{e_i\otimes f_j, e_l\otimes f_m}\Big)$$

and

$$R_1\Big(\sum_{jknm}\beta_{jk,mn}\Theta_{f_j\otimes g_k, f_m\otimes g_n}\Big) = \sum_{jm}\Big(\sum_k \beta_{jk,mk}\Theta_{f_j,f_m}\Big)$$

This shows that $R_2 = id_{M_k} \otimes R_1$. It follows now from the definitions that $\varphi_{\underline{t}}^{(k)} = id_{M_k} \otimes \varphi_{\underline{t}}$. $\square$

Theorem 3.3.9 *For any continuous tensor product T of type I, $\mathcal{A}(T)$ is nuclear.*

<u>Proof:</u> Let $x_k \in M_k(\mathcal{S}(T))$, $\mathcal{S}$ as in Rem.3.1.10. It is easy to see that there is a sequence $\underline{t}(l)$ of finer and finer partitions $a < t_1(l) < \ldots < t_{n_l}(l) < b$ s.t. $\varphi_{\underline{t}(l)}^{(k)}(\omega_k)(x_k) \xrightarrow{l\to\infty} \omega_k(x_k)$ whenever $\omega_k \in M_k(\mathcal{A}(T))^*$.

For any sequence $a < t_1 < \ldots < t_n < b$ we have onb's $(e_{m_0}(t_1) \otimes f_{m_1}(t_1, t_2) \otimes \ldots \otimes f_{m_n}(t_n, b))$ of $\mathcal{H}$ as in the proof of Rem.3.2.14 and to each finite sequence $M = (m_0, \ldots, m_n)$ let P_i be the finite rank projection onto $[e_{k_0}(t_0) \otimes \ldots \otimes f_{k_{i-1}}(t_{i-1}, t_i) | k_r < m_r,\ r = 0, \ldots, i-1] \subseteq \mathcal{H}_{t_i}$. Define

$$\varphi_{\underline{t},M}((T_t)) = \begin{cases} R_t^{t_i}(P_i T_{t_i} P_i) & \text{if } i \in (t_{i-1}, t_i],\ i = 1, \ldots, n \\ 0 & \text{otherwise} \end{cases}$$

$\varphi_{\underline{t},M}((T_t))$ is still a positive contraction. Replacing the sequence P_i by the sequence of projections $\mathbf{1}_{\mathbb{C}^k} \otimes P_i$ in $\mathbb{C}^k \otimes \mathcal{H}_{t_i}$, we obtain $\varphi_{\underline{t},M}^{(k)} = id_{M_k} \otimes \varphi_{\underline{t},M}$ and this map is positive. Thus $\varphi_{\underline{t},M}$ is completely positive and of finite rank. Choosing an appropriate sequence $M(l)$, we still have $\varphi_{\underline{t}(l),M(l)}^{(k)}(\omega_k)(x_k) \xrightarrow{l\to\infty} \omega_k(x_k)$. It follows from [Lan 73, Thm.3.6] that $\mathcal{A}(T)$ is nuclear. $\square$

3.3.2 General Diagonal Morphisms

To obtain all automorphisms of $\mathcal{A}(T)$ one only needs to find the ideal preserving ones and a single one s.t. the corresponding map on the ideal space is a given $\sigma : [a,b] \to [a,b]$. Looking at the algebra $C_0(0,1]$, one would expect that σ could be any monotone invertible map. However, it turns out that in the diagonal case at least it has certain smoothness properties.

Similarly, any isomorphism between $\mathcal{A}(T)$ and $\mathcal{A}(T')$ (or $\mathcal{A}(T) \otimes \mathcal{K}$ and $\mathcal{A}(T') \otimes \mathcal{K}$) gives a monotone invertible map $\sigma : I \to I'$. But there are further restrictions in this case.

We consider now two type I continuous tensor decompositions T and T' given by direct integrals $\int_I^\oplus \mathcal{K}_t d\mu$ and $\int_{I'}^\oplus \mathcal{L}_{t'} d\nu$ s.t. $\tau : I' \to I$ implements a change of measure (c.f. Appendix sec.B.1). Let with $\tau := \sigma^{-1}$

$$V_\tau(f) = \sqrt{\tfrac{d\mu \circ \tau}{d\nu}} f \circ \tau \ , \ f \in \int_I^\oplus \mathcal{K}_t d\mu$$

Thus $exp(f) \mapsto exp(V_\tau(f))$ is a unitary u. Put $\alpha = Ad(u)$ in the regular representations. (c.f. Lemma 3.1.12)

Remark 3.3.10 α *defines an isomorphism of* $\mathcal{A}(T)$ *onto* $\mathcal{A}(T')$ *s.t.* $\alpha(\mathcal{A}^s) = \mathcal{A}^{\sigma(s)}$.

Proof: Compare with the proof of 3.1.12. □

$u : \mathcal{F}^s(\int_I^\oplus \mathcal{K}_t d\mu) \to \mathcal{F}^s(\int_{I'}^\oplus \mathcal{L}_{t'} d\nu)$ unitary is called factorizable w.r.t. $\sigma : [a,b] \to [a',b']$ monotone invertible with $\sigma(a) = a'$ if for $a < t_1 < \ldots < t_k < b$ we have

$$u_{[t_i,t_{i+1})} : \mathcal{H}_{[t_i,t_{i+1})} \to \mathcal{H}'_{[\sigma(t_i),\sigma(t_{i+1}))}$$

$(t_0 = a, t_{k+1} = b)$ s.t.

$$u = u_{(a,t_1)} \otimes \ldots \otimes u_{[t_k,b)}$$

Assertions about factorizable maps as in [Gui 72] generalize immediately to σ-factorizable maps.

Lemma 3.3.11 *Let* $\sigma : [a,b] \to [a',b']$ *monotone invertible be given and* u *factorizable w.r.t.* σ *as above. Then* $\tau := \sigma^{-1} : I' \to I$ *defines a change of measure and* u *is of the form*

$$u(exp(f)) = e^{-\|h\|^2/2 - \langle h, V_\tau Df \rangle} exp(V_\tau Df + h)$$

for some diagonal unitary D, *some* $h \in \int_I^\oplus \mathcal{K}_t d\mu$ *and* V_τ *as above.*

Proof: First of all, u preserves factorizable vectors and hence is of the form $u = W(V, h, c)$, $V : \int_I^\oplus \mathcal{K}_t d\mu \to \int_{I'}^\oplus \mathcal{L}_{t'} d\nu$ unitary, $h \in \int_{I'}^\oplus \mathcal{L}_{t'} d\nu$, $c \in S^1$ using Lemma 2.1.15. For $\Delta \in \mathcal{B}_0(I)$ we have $supp(Vf) \subseteq \sigma(\Delta)$ if $supp\, f \subseteq \Delta$ as may be seen as follows: Let $supp\, f \subseteq \Delta$ then

$$u(exp(f)) = \underbrace{u_\Delta(exp(f|\Delta))}_{\in \mathcal{H}'_{\sigma\Delta}} \otimes \underbrace{u_{\Delta^c}(exp(0))}_{\in \mathcal{H}'_{\sigma\Delta^c}}$$

For arbitrary $f_1 \in \int_I^\oplus \mathcal{K}_t d\mu$ we have

$$u(exp(f_1)) = k\, exp(Vf_1|\sigma\Delta + h_0) \otimes exp(Vf_1|\sigma\Delta^c + h_1)$$

for some $h_0 \in \int_{\sigma\Delta}^\oplus \mathcal{L}_{t'} d\nu$, $h_1 \in \int_{\sigma\Delta^c}^\oplus \mathcal{L}_{t'} d\nu$ and $k \in \mathbb{C}$.

Putting $f_1 = 0$, we get $u(exp(0)) = k'\, exp(h) = k''\, exp(h_0) \otimes exp(h_1)$ for some $k', k'' \in \mathbb{C}$ and obtain $h_0 = h|\sigma\Delta$, $h_1 = h|\sigma\Delta^c$.

Putting $f_1 = f$ we get $Vf_1|\sigma\Delta^c + h|\sigma\Delta^c = h_1 = h|\sigma\Delta^c$. Hence $supp(Vf_1) \subseteq \sigma\Delta$. This shows $V = V_\tau D$ for some diagonal operator D by Appendix Rem.9. □

We say for the moment that $\tau : I' \to I$ defines almost a change of measure if $\tau|(a', t') \to (a, \tau(t'))$ defines a change of measure on the restrictions whenever $a' < t' < b'$.

Proposition 3.3.12 *There exist a diagonal isomorphism between $\mathcal{A}(T)$ and $\mathcal{A}(T')$ iff $\int_I^\oplus \mathcal{K}_t d\mu$ and $\int_{I'}^\oplus \mathcal{L}_{t'} d\nu$ differ almost by a change of measure.*

Proof: Let $\varphi : \mathcal{A}(T) \to \mathcal{A}(T')$ be a diagonal isomorphism. Let $\sigma : I \to I'$ be the monotone invertible map defined by $\varphi(\mathcal{A}^s(T)) = \mathcal{A}^{\sigma(s)}(T')$. The locally pure states on $\mathcal{A}(T)$ are in 1-1-correspondence with families of rays $[\xi_t]$, $\xi_t \in \mathcal{H}_t$, $\|\xi_t\| = 1$, $\xi_t = \xi_s \otimes \xi_{[s,t)}$ for $a < s < t$. Now we see that ξ_t is a multiple of an exponential vector.

An argument like in the proof of Thm.3.3.6.(i) shows that for each $a < t < b$ we obtain σ-factorizable maps $u_t : \mathcal{H}_t \to \mathcal{H}'_{\sigma(t)}$.

It now follows from Lemma 3.3.11 that the two direct integrals differ almost by a change of measure. The converse is obvious using Rem.3.3.10. □

In effect the multiplicity in $\int_I^\oplus \mathcal{K}_t d\mu$ has some relevance for $\mathcal{A}(T)$.

3.3.3 Generation by Cones

A. Factorizable Vectors and Spectral Resolutions

By Prop.3.2.15 we can think of $\mathcal{A}(T)$ as concrete C^*-algebra on $\mathcal{H}$. For $f \in \int_I^\oplus \mathcal{K}_t d\mu$ observe that

$$t \mapsto e^{-\|f_t\|^2} \Theta_{exp(f_t),exp(f_t)} \otimes \mathbf{1}$$

is a decreasing family of projections in $\mathcal{H} = \mathcal{F}^s(\int_I^\oplus \mathcal{K}_t d\mu)$. Moreover, this family is left (and right) continuous and therefore a spectral resolution in the sense of Appendix Def.11.

Remark 3.3.13 *Each factorizable vector in $\mathcal{H}$ induces an inclusion of $C_0(a,b]$ into $\mathcal{A}(T)$.*

Proof: We only need to show that the spectral measure E corresponding to the resolution above has full support (i.e. is equal to I). By conjugation with the Weyl unitary $W(1,-f)$, we may assume f equal to 0. But then $E(\Lambda) \geq F(\Lambda)$ for $\Lambda \subseteq I$ any Borel set where F is the projection valued measure given by multiplication with χ_Λ in $\int_I^\oplus \mathcal{K}_t d\mu \subseteq \mathcal{F}^s(\int_I^\oplus \mathcal{K}_t d\mu)$. F has full support. $\square$

We call algebras isomorphic to $C_0(0,1]$ cones. $\mathcal{A}(T)$ may be generated by all the ones of the above form but it turns out that one often only needs a much smaller number of them. In the rest of this chapter we consider only the algebras $\mathcal{A}_n$. It turns out that they can be generated by $n+1$ cones. In the general case where T is of type I, things are more complicated. There don't seem to be any canonical cones if T is type III.

We introduce some notation: For $(x,y) \subseteq (0,1)$ let

$$\epsilon_i(x,y) = \begin{cases} e^{-(y-x)/2} exp(\chi_{(x,y)} \otimes e_i) & \text{if } i = 1,\ldots,n \\ exp(0|(x,y)) & \text{if } i = 0 \end{cases}$$

where $\chi_{(x,y)} \otimes e_i$ means the function on the interval (x,y) identically equal to the i-th unit vector e_i in $\mathbb{C}^n$.

ALGEBRAS ASSOCIATED TO CONTINUOUS T.P.

Let $\epsilon_{ij}(x,y) := \Theta_{\epsilon_i(x,y),\epsilon_j(x,y)}$ which is a rank 1 operator in $\mathcal{H}_{(x,y)}$. Because the $\epsilon_i(x,y)$ are factorizable they never become quite orthogonal. Therefore the ϵ_{ij} don't behave like matrix units.

In what follows we restrict again to $I = (0,1)$ (and $\mu = dt$). Everything holds for any I with obvious modifications. (Note that if $I = (-\infty, b)$, we have to replace the constant function $\chi \otimes e_i$ by $f \otimes e_i$, $f \in L^2(I)$)

Lemma 3.3.14 *For $s, t > 0$ the matrix (γ_{ij}) given by*

$$(\gamma_{ij}) = \begin{pmatrix} 0 & 1/2 & \cdots & \cdots & 1/2 \\ 1/2 & 0 & 1 & \cdots & 1 \\ \vdots & 1 & 0 & \cdots & 1 \\ \vdots & \vdots & \vdots & \ddots & \vdots \\ 1/2 & 1 & \cdots & \cdots & 0 \end{pmatrix}$$

is conditionally positive definite (c.f. Appendix C) and

$$\epsilon_{ij}(0,t)\epsilon_{kl}(0,s) = \begin{cases} e^{-\gamma_{jk}t}\epsilon_{il}(0,t) \otimes \epsilon_{kl}(t,s) & \text{if } t < s \\ \\ e^{-\gamma_{jk}s}\epsilon_{il}(0,s) \otimes \epsilon_{ij}(s,t) & \text{otherwise} \end{cases}$$

<u>Proof:</u> Direct computation. □

Now for each $i = 0, \ldots, n$ we have the spectral resolution

$$t \mapsto \epsilon_{ii}(0,t) \otimes \mathbf{1} =: p_t^i$$

Let $\mathcal{C}_n$ be the closed subalgebra generated by $\{\int_0^1 dt f(t) p_t^i \mid f \in L^1(I), i = 0, \ldots, n\}$. Consider the simplex domain $S_p := \{(t_1, \ldots, t_p) \mid 0 < t_1 < \ldots < t_p < 1\}$ equiped with the Lebesgue measure.

Remark 3.3.15 *For $i_1, \ldots, i_p \in \{0, \ldots, n\}$ and $f \in L^1(S_p)$*

$$x := \int_0^1 dt_p \int_0^{t_p} dt_{p-1} \ldots \int_0^{t_2} dt_1 f(t_1, \ldots, t_p) p_{t_1}^{i_1} \ldots p_{t_p}^{i_p}$$

lies in $\mathcal{C}_n$.

<u>Proof:</u> For $f_1, \ldots, f_p \in L^1(I)$ and $i_1, \ldots, i_p \in \{0, \ldots, n\}$

$$\int_0^1 dt_p \int_0^1 dt_{p-1} \ldots \int_0^1 dt_1 f_1(t_1) \ldots f_p(t_p) p_{t_1}^{i_1} \ldots p_{tp}^{i_p}$$

lies in $\mathcal{C}_n$ and hence for each $f \in L^1(I^p)$

$$\int_0^1 dt_p \int_0^1 dt_{p-1} \ldots \int_0^1 dt_1 f(t_1, \ldots, t_p) p_{t_1}^{i_1} \ldots p_{t_p}^{i_p}$$

lies in $\mathcal{C}_n$ by approximating f by sums of products (one can use for instance continuous functions of compact support and Stone-Weierstrass) and $L^1(S_p) \subseteq L^1(I^p)$. □

Lemma 3.3.16 *For each normal state φ (w.r.t. π_0) there exists $x \in \mathcal{C}_n$ s.t. $\varphi(xx^*) > 0$.*

<u>Proof:</u> We may assume φ to be a vector state $\varphi = \langle \xi, \cdot \xi \rangle$. By 2.1.19 the set $S := \{\epsilon_{i_1}(0, t_1) \otimes \ldots \otimes \epsilon_{i_p}(t_{p-1}, 1) | i_1, \ldots, i_p \in \{0, \ldots, n\}, 0 < t_1 < \ldots < t_{p-1} < 1\}$ is spanning. We can find $0 < t_1 < \ldots < t_{p-1} < 1$, $i_1, \ldots, i_p \in \{0, \ldots, n\}$ s.t.

$$\langle \epsilon_{i_1}(0, t_1) \otimes \ldots \otimes \epsilon_{i_p}(t_{p-1}, 1), \xi \rangle \neq 0$$

By continuity in the parameters (which follows from the continuity of exp), we have

$$\langle \epsilon_{i_1}(0, s_1) \otimes \ldots \otimes \epsilon_{i_p}(s_{p-1}, s_p) \otimes \epsilon_0(s_p, 1), \xi \rangle \neq 0$$

provided $|s_i - t_i| < \varepsilon$, $1 - s_p < \varepsilon$ with a sufficiently small $\varepsilon > 0$. Extending $\epsilon_0(t, 1)$ to an onb of $\mathcal{H}^t$, it follows that

$$\langle \xi, (\epsilon_{i_1 i_1}(0, s_1) \otimes \ldots \otimes \epsilon_{i_p i_p}(s_{p-1}, s_p) \otimes \epsilon_{00}(s_p, t) \otimes \mathbf{1}) \xi \rangle$$

$$\geq |\langle \xi, \epsilon_{i_1}(0, s_1) \otimes \ldots \otimes \epsilon_{i_p}(s_{p-1}, s_p) \otimes \epsilon_0(s_p, 1) \rangle|^2 > 0$$

if $s_p < t < 1$. Therefore

$$Re \langle \xi, (\Theta_{\epsilon_{i_1}(0, s_1) \otimes \ldots \otimes \epsilon_{i_p}(s_{p-1}, s_p) \otimes \epsilon_0(s_p, t), \epsilon_{i_1}(0, s_1') \otimes \ldots \otimes \epsilon_{i_p}(s_{p-1}', s_p') \otimes \epsilon_0(s_p', t)} \otimes \mathbf{1}) \xi \rangle$$

is strictly positive if $|t_i - s_i|, |t_i - s_i'| < \varepsilon$, $\varepsilon < 1 - s_p$, $1 - s_p' < 2\varepsilon$, $t > 1 - \varepsilon$ for a possibly smaller $\varepsilon > 0$. Put $t_p = 1 - \varepsilon$ and define the function

$$f_p^\varepsilon(s_1, \ldots, s_p, s) = \begin{cases} \frac{1}{\varepsilon^{p+1}} & \text{if } 0 \leq t_i - s_i < \varepsilon, s \in (1-\varepsilon, 1) \\ 0 & \text{otherwise} \end{cases}$$

ALGEBRAS ASSOCIATED TO CONTINUOUS T.P.

Let
$$x_\varepsilon = \int_0^1 ds \int_0^s ds_p \ldots \int_0^{s_2} ds_1 \, f_p^\varepsilon(s_1, \ldots, s_p, s) \, p_{s_1}^{i_1} \ldots p_{s_p}^{i_p} p_s^0$$

Then
$$\langle \xi, x_\varepsilon x_\varepsilon^* \xi \rangle = Re\langle \xi, x_\varepsilon x_\varepsilon^* \xi \rangle > 0$$

which concludes the proof. □

Proposition 3.3.17 $\mathcal{C}_n = \mathcal{A}_n$ i.e. $\mathcal{A}_n$ is generated by $n+1$ algebras isomorphic to $C_0(0,1]$.

<u>Proof:</u> For $t \in (0,1)$ let $S_t = \{\epsilon_{i_1}(0,t_1) \otimes \ldots \otimes \epsilon_{i_{p+1}}(t_p,t) | i_1, \ldots, i_{p+1} \in \{0, \ldots, n\}, 0 < t_1 < \ldots < t_p < t < 1\}$. Then $[S_t] = \mathcal{H}_t$ by 2.1.19. Let $i_1, \ldots, i_{p+1} \in \{0, \ldots, n\}, t, t_i \in \mathbb{Q}, 0 < t_1 < \ldots < t_p < t < 1$ and $\varepsilon > 0$ also rational and sufficiently small. Define

$$g_{p,t}^\varepsilon(s_1, \ldots, s_p, s) = \begin{cases} \frac{1}{\varepsilon^{p+1}} & \text{if } 0 \leq t_i - s_i < \varepsilon, s \in (t, t+\varepsilon) \\ 0 & \text{otherwise} \end{cases}$$

Let x_ε an x_ε' be defined by the same integral as in the foregoing proof with f_p^ε replaced by $g_{p,t}^\varepsilon$ and $g_{p',t'}^\varepsilon$. Consider the set R of elements in $\mathcal{C}_n$ of the form $x_\varepsilon x_\varepsilon'^*$ with arbitrary $g_{p,t}^\varepsilon, g_{p',t'}^\varepsilon$ as above s.t. $|t - t'| > \varepsilon$. This set is countable and using 2.1.19 one can see that its span over $\mathbb{Q}(i)$ is generating in the sense of Appendix Prop.7 as a set of L^1-sections. On the other hand, a little calculation shows that multiplication of the elements in R considered as sections by $C_c(0,1)$-functions does not lead out of $\mathcal{C}_n$. It follows from Appendix Prop.7 that $\mathcal{C}_n$ contains $L^1(\mathcal{T}_n)$ hence equals $\mathcal{A}_n$. □

3.3.4 Pedersen Ideal and Infiniteness

We denote our generating canonical cones by $C^i := C^*(\{\int_0^1 dt \, f(t) p_t^i \, | f \in L^1(0,1)\})$, $i = 0, \ldots, n$. In each ideal $\mathcal{A}_n^s \cong \mathcal{K} \otimes \mathcal{A}_n$ consider the subalgebra

$$\check{\mathcal{A}}_n^s := \mathcal{F} \odot \mathcal{A}_n$$

where $\mathcal{F}$ are the finite rank operators and $\odot$ is the algebraic tensor product.

Lemma 3.3.18 *If $n < \infty$, then $\check{\mathcal{A}}_n^s$ is a nonclosed ideal in $\mathcal{A}_n$ s.t. $\check{\mathcal{A}}_n^{s_1} \subseteq \check{\mathcal{A}}_n^{s_2}$ if $s_1 \geq s_2$ and the union (i.e. the algebraic inductive limit)*

$$\mathcal{P}(\mathcal{A}_n) = \bigcup_{s \in I} \check{\mathcal{A}}_n^s$$

is the Pedersen ideal of $\mathcal{A}_n$.

Proof: In the regular representation we have by Prop.3.2.17 $\mathcal{A}_n = \mathcal{K} \otimes \mathcal{A}_n \oplus \mathcal{A}_n \otimes \mathbf{1}$ (omitting s). Hence $\mathcal{F} \odot \mathcal{A}_n$ is clearly an ideal in $\mathcal{A}_n$ and therefore $\mathcal{P}(\mathcal{A}_n) \subseteq \bigcup_{s \in I} \check{\mathcal{A}}_n^s$. On the other hand, $\check{\mathcal{A}}_n^s = \mathcal{F} \odot \mathcal{A}_n \subseteq \mathcal{P}(\mathcal{A}_n)$ for the following reason: $\mathcal{P}(\mathcal{A}_n)$ contains the set $C_\varepsilon^i := C^*(\{\int_0^{1-\varepsilon} da\ f(a)\epsilon_{ii}(0, \varepsilon + a) | f \in L^1(0, 1-\varepsilon)\})$ for each $i \in \{0, \ldots, n\}$. (These are the functions in $C^i \cong C_0(0,1]$ having support in $[\varepsilon, 1]$). Let $\xi_\varepsilon \in \mathcal{H}_\varepsilon$ with $\|\xi_\varepsilon\| = 1$ and for $0 < \varepsilon < s < 1$

$$x := \int_\varepsilon^1 ds\ \Theta_{\xi_\varepsilon, \epsilon_0(0,\varepsilon)} \otimes \epsilon_{ii}(\varepsilon, s) \otimes \mathbf{1} \in \mathcal{A}_n$$

One computes

$$x \left(\int_\varepsilon^1 ds\ \epsilon_{00}(0,\varepsilon) \otimes \epsilon_{ii}(\varepsilon, s) \otimes \mathbf{1} \right) x^* = \Theta_{\xi_\varepsilon, \xi_\varepsilon} \otimes \left(\int_0^{1-\varepsilon} da\ 3a^2\ \epsilon_{ii}(\varepsilon, \varepsilon + a) \otimes \mathbf{1} \right)$$

Hence $\mathcal{P}(\mathcal{A}_n)$ contains $\Theta_{\xi_\varepsilon, \xi_\varepsilon} \otimes C_\varepsilon^i$ (somewhat informal). Because $\mathcal{P}(\mathcal{A}_n)$ is closed under the formation of finitely generated C^*-subalgebras ([Ped 79 5.6.2]) we obtain $p \otimes \mathcal{A}_n(\varepsilon, 1) \subseteq \mathcal{P}(\mathcal{A}_n)$ for each minimal projection $p \in \mathcal{K}$, so $\mathcal{F} \odot \mathcal{A}_n(\varepsilon, 1) \subseteq \mathcal{P}(\mathcal{A}_n)$ for each $\varepsilon > 0$. □

Lemma 3.3.19 *There is no bounded trace on $\mathcal{A}_n$.*

Proof: Suppose ψ is a tracial state and (T_t) as in Cor.3.2.10.(ii). There is $t_0 \in I$ s.t. $T_t \neq 0$ for $t < t_0$. In particular, the corresponding GNS-representation π_ψ is faithful on $\mathcal{A}_n(0, t_0)$ and $\psi | \mathcal{A}_n(0, t_0)$ is a faithful trace. Restricting ψ to $C^*(\{\int_\varepsilon^{t_0} dt\ f(t) K_\varepsilon \otimes \epsilon_{00}(\varepsilon, t) \otimes \mathbf{1} | f \in L^1(\varepsilon, t_0), K_\varepsilon \in \mathcal{K}_\varepsilon\}) \subseteq \mathcal{A}_n$, we get a nonzero bounded trace on $\mathcal{K}$ using an approximate unit in $C_0(\varepsilon, t_0]$. □

Corollary 3.3.20 *There is no lower semicontinuous densely defined trace on $\mathcal{A}_n$.*

Proof: By [Cu 82b] each such trace is defined on $\mathcal{P}(\mathcal{A}_n)$ hence also on $\check{\mathcal{A}}_n^\varepsilon$, in particular on $p \otimes \mathcal{A}_n$ with p a minimal projection and thus defines a bounded trace on $\mathcal{A}_n$, contradicting the foregoing Lemma. □

3.3.5 The Canonical Automorphic and Endomorphic Actions on $\mathcal{A}_n$

Taking $I = \mathbb{R}$ and $I = \mathbb{R}_+$, we get automatically auto- and endomorphic actions of $\mathbb{R}$ and $\mathbb{R}_+$ on $\mathcal{A}_n$.

Remark 3.3.21 *Let $(\alpha_t)_{t\in\mathbb{R}}$ $((\beta_t)_{t\in\mathbb{R}_+})$ be a group of automorphisms (a semi-group of endomorphisms) on a C^*-algebra A. Suppose there is $\mathcal{D} \subseteq A$ norm-dense s.t. for each $x \in \mathcal{D}$ the map $t \mapsto \alpha_t(x)$ ($t \mapsto \beta_t(x)$) is norm continuous. Then (α_t) $((\beta_t))$ is strongly continuous, i.e. continuity holds for each $x \in A$.*

<u>Proof:</u> Consider first the automorphic case. Let $a \in A$ and $\varepsilon > 0$. We have to show that there exists $\delta > 0$ s.t. $\forall t \in (-\delta, \delta) : \|\alpha_t(a) - a\| < \varepsilon$. Take $x \in \mathcal{D}$ s.t. $\|x - a\| < \frac{\varepsilon}{3}$ and $\delta > 0$ s.t. $\|\alpha_t(x) - x\| < \frac{\varepsilon}{3}$ $\forall t \in (-\delta, \delta)$. Then $\|\alpha_t(a) - a\| \leq \|\alpha_t(a) - \alpha_t(x)\| + \|\alpha_t(x) - x\| + \|x - a\| < \varepsilon$ because $\|\alpha_t(a-x)\| = \|a - x\|$. The proof of the endomorphic case is similar, one has to replace $0 \in \mathbb{R}$ by $t_0 \in \mathbb{R}_+$. □

Proposition 3.3.22 *(i) Let $\mathcal{A}_n = \mathcal{A}_n(\mathbb{R})$ be given over $I = \mathbb{R}$ in the regular representation and let $\alpha_t := Ad(exp(S_t))$, $S_t : L^2(\mathbb{R}, \mathbb{C}^n) \to L^2(\mathbb{R}, \mathbb{C}^n)$ be the two sided shift. Then α_t is a strongly continuous action of $\mathbb{R}$ on $\mathcal{A}_n$.*

(ii) Let $\mathcal{A}_n = \mathcal{A}_n(\mathbb{R}_+)$ be given over $\mathbb{R}_+$ in the regular representation and define $\rho_t(X) = exp(T_t) \, X \, exp(T_t)^*$ for $X \in \mathcal{B}(\mathcal{F}^s(L^2(\mathbb{R}_+, \mathbb{C}^n)))$. Then ρ_t leaves $\pi_0(\mathcal{A}_n)$ invariant and defines a strongly continuous semigroup of endomorphisms on $\mathcal{A}_n$ s.t. $\rho_t(\mathcal{A}_n) \neq \mathcal{A}_n$ for each positive t.

<u>Proof:</u> (i) α_t is nothing else but $Ad(u)$ considered in Rem.3.3.10 where $\sigma : [-\infty, \infty] \to [-\infty, \infty]$, $\sigma(x) = x + t$ for $x \in \mathbb{R}$ and $\sigma(-\infty) = -\infty$, $\sigma(\infty) = \infty$. We only have to check strong continuity. Using Rem.3.3.21, it suffices to do that for $x \in \mathcal{S}_n$. Let $x = \int dt \, \varphi(t) \Theta_{exp(f_t), exp(g_t)}$, $\varphi \in C_c(\mathbb{R})$, $f, g \in L^2(\mathbb{R}, \mathbb{C}^n)$. Then for $r \in \mathbb{R}$

$$\alpha_r(x) = \int dt \, \varphi(t) \, \Theta_{exp(S_r f_t), exp(S_r g_t)}$$

where by $S_r f_t$ we mean $f_t = f|(-\infty, t)$ extended by 0, shifted by r and restricted to $(-\infty, r + t)$ (i.e. $exp(S_t)$ as a unitary identifying

$\mathcal{F}^s(L^2((-\infty,t),\mathbb{C}^n))$ and $\mathcal{F}^s(L^2((-\infty,t+r),\mathbb{C}^n)))$. Now

$$\|\alpha_r(x) - x\| \leq \int dt \, \|\varphi(t-r)\,\Theta_{exp(S_r f_{t-r}),exp(S_r g_{t-r})} - \varphi(t)\,\Theta_{exp(f_t),exp(g_t)}\|$$

$$\leq M \int_{-t_0}^{t_0} dt \, \big[\|exp(S_r f_{t-r}) - exp(f_t)\|$$
$$+ \quad \|exp(S_r g_{t-r}) - exp(g_t)\|\big]$$

where M is a constant depending on φ ($M \leq 2\|\varphi\|_\infty$) and $t_0 > 0$ is s.t. $supp(\varphi) \cup (supp(\varphi) + r) \subseteq [-t_0, t_0]$. With a suitable $c > 0$ we have

$$\|\alpha_r(x) - x\| \leq M \int_{-t_0}^{t_0} dt \, c \, \big[\|S_r f_{t-r} - f_t\| + \|S_r g_{t-r} - g_t\|\big]$$

by continuity of exp as a nonlinear map and finiteness of t_0. But $\|S_r f_{t-r} - f_t\| \leq \|S_r f_{t-r} - f_{t-r}\| + \|f|(t-r,t)\|$, f_{t-r} extended by 0. Both terms tend to 0 for $r \to 0$.

(ii): In case of ρ_t remark that if we identify $\mathcal{A}_n(\mathbb{R}_+)$ with the subalgebra $\epsilon_{00}(-\infty, 0) \otimes \mathcal{A}_n(\mathbb{R}_+) \subseteq \mathcal{A}_n(\mathbb{R})$, then ρ_t is the restriction of α_t to this subalgebra. □

There is also the possibility to define this action on $L^1(T_n)$ which will be taken up in Ch.4.

3.4 Homotopy Invariants

3.4.1 K-Theory

We call any functor F from C^*-algebras to sets a representable homotopy functor if it is of the form

$$F(A) = [\mathcal{F}, A]$$

where $\mathcal{F}$ is a certain C^*-algebra and $[A, B]$ denotes the set of homotopy classes of homomorphisms between the C^*-algebras A and B. F is called stable if $F(A) = [\mathcal{F}, A \otimes \mathcal{K}]$. If F has values in abelian groups, we can say what halfexactness means.

Proposition 3.4.1 *For each representable (stable) halfexact homotopy functor F from C^*-algebras to abelian groups and $n \in \mathbb{N}$ we have*

$$F(\mathcal{A}_n) = 0$$

Proof: Let F be stable, $F(A) = [\mathcal{F}, A \otimes \mathcal{K}]$, $\mathcal{F}$ a C^*-algebra. Represent $\mathcal{A}_n$ as $\mathcal{A}_n(\mathbb{R}_+)$. We have the split exact sequence (c.f.3.2.17)

$$0 \to \mathcal{A}_n^1 \to \mathcal{A}_n \overset{\leftarrow}{\to} \mathcal{A}_n(0,1) \to 0$$

Intersection with $\mathcal{A}_n^\varepsilon$ for some $0 < \varepsilon < 1$ or tensoring with $\mathcal{K}$ makes everything stable and split:

$$0 \to \mathcal{A}_n^1 \to \mathcal{A}_n^\varepsilon \overset{\leftarrow}{\to} \mathcal{A}_n(\varepsilon,1) \otimes \mathcal{K} \to 0$$

Applying F we obtain by half-exactness

$$F(\mathcal{A}_n^1) \to F(\mathcal{A}_n^\varepsilon) \overset{\leftarrow}{\to} F(\mathcal{A}_n(\varepsilon,1)) \to 0$$

But any $\varphi \in Hom(\mathcal{F}, \mathcal{A}_n^\varepsilon)$ is homotopic to a homomorphism with image in $\mathcal{A}_n^1$ by the homotopy

$$[0,1] \ni \lambda \mapsto \varphi \circ \rho_{\lambda(1-\varepsilon)}$$

Hence $F(\mathcal{A}_n^1) \to F(\mathcal{A}_n^\varepsilon)$ is surjective and therefore $F(\mathcal{A}_n(\varepsilon,1)) = F(\mathcal{A}_n) = 0$. In the unstable case we do the same without stabilizing. Note that this homotopy also shows that there are no nonzero projections in $\mathcal{A}_n$ and $\mathcal{A}_n \otimes \mathcal{K}$.
□

Corollary 3.4.2 $\mathcal{A}_n$ *is KK-contractible and stably projectionless.*

Proof: KK-contractible means that $KK(\mathcal{A}_n, B) = 0 = KK(B, \mathcal{A}_n) = 0$ for any σ-unital C^*-algebra B. By the Kasparov product it is enough to have $KK(\mathcal{A}_n, \mathcal{A}_n) = 0$. But this is true because $KK(A,B) = [qA, B \otimes \mathcal{K}]$ by [Cu 92]. □

3.4.2 The Homotopy Type of the Automorphism Group

Consider $Aut(\mathcal{A}_n)$ with the strong topology (i.e. a subbasis of the topology is given by the neighborhoods $U(\alpha_0, x, \varepsilon) := \{\alpha \mid \|\alpha(x) - \alpha_0(x)\| < \varepsilon\}$).

Proposition 3.4.3 $Aut(\mathcal{A}_n)$ *is contractible.*

<u>Proof:</u> Let $I = \mathbb{R}_+$. For each $t \in I$ we have the splitting
$$\mathcal{A}_n = \mathcal{A}_n(0,t) \oplus \mathcal{A}_n^t = (\mathcal{A}_n \otimes \mathbf{1}) \oplus (\mathcal{K} \otimes \mathcal{A}_n)$$
For $\alpha \in Aut(\mathcal{A}_n)$ define a new automorphism $id_t \otimes \alpha$ by
$$(id_t \otimes \alpha)(x \otimes y) = \begin{cases} x \otimes y & \text{if } x \otimes y \in \mathcal{A}_n \otimes \mathbf{1} \\ x \otimes \alpha(y) & \text{if } x \otimes y \in \mathcal{K} \otimes \mathcal{A}_n, \end{cases}$$
where we have to fix an identification of $\mathcal{A}_n(0,\infty)$ and $\mathcal{A}_n(t,\infty)$ for each $t > 0$ as follows:

In the representation space of the regular representation of $\mathcal{A}_n(0,\infty)$ there is the strongly continuous semigroup $U_0(t) := exp(T_t)$ with T_t the onesided shift on $L^2(\mathbb{R}_+, \mathbb{C}^n)$. Now $\mathcal{A}_n(t,\infty)$ is not a subalgebra of $\mathcal{A}_n(0,\infty)$ but may be identified with $\epsilon_{00}(0,t) \otimes \mathcal{A}_n(t,\infty) = p_t \otimes \mathcal{A}_n(t,\infty)$. Then $U_0(t)^*(p_t \otimes \mathcal{A}_n(t,\infty))U_0(t) = \mathcal{A}_n(0,\infty)$ and it defines an isomorphism between the two algebras. In order to see that $id_t \otimes \alpha$ defines an automorphism we only have to remark that for $x \otimes \mathbf{1} \in \mathcal{A}_n \otimes \mathbf{1}$ and $y \otimes z \in \mathcal{K} \otimes \mathcal{A}_n$ we have

$$\begin{aligned}((id_t \otimes \alpha)(x \otimes \mathbf{1}))((id_t \otimes \alpha)(y \otimes z)) &= (x \otimes \mathbf{1})(y \otimes \alpha(z)) \\ &= xy \otimes \alpha(z) = (id_t \otimes \alpha)(xy \otimes z)\end{aligned}$$

(All other cases are clear). For $x = \int ds\, \varphi(s) \Theta_{exp(f_s), exp(g_s)} \in \mathcal{S}_n$ (integral in the regular representation) we have

$$\begin{aligned}(id_t \otimes \alpha)(x) &= \int_0^t ds\, \varphi(s)\, \Theta_{exp(f_s), exp(g_s)} + \\ &\quad + p_t \left(\alpha \left(\int_t^\infty ds\, \varphi(s)\, \Theta_{exp(S_{-t}f_{[t,s)}), exp(S_{-t}g_{[t,s)})} \right) \right)\end{aligned}$$

and this is continuous in t, the second term being zero for $t > supp(\varphi)$. Now we argue as in Rem.3.3.21 : For $a \in \mathcal{A}_n$ take $x \in \mathcal{S}_n$ s.t. $\|x - a\| < \varepsilon/3$ and $\delta > 0$ s.t. $\forall |t - t_0| < \delta : \|(id_t \otimes \alpha)(x) - (id_{t_0} \otimes \alpha)(x)\| < \frac{\varepsilon}{3}$. Then

$$\|(id_t \otimes \alpha)(a) - (id_{t_0} \otimes \alpha)(a)\| = \|(id_t \otimes \alpha)(a) - (id_t \otimes \alpha)(x)\| +$$
$$+ \;\|(id_t \otimes \alpha)(x) - (id_{t_0} \otimes \alpha)(x)\| + \|(id_{t_0} \otimes \alpha)(x) - (id_{t_0} \otimes \alpha)(a)\| < \varepsilon$$

Hence $t \mapsto (id_t \otimes \alpha)(a)$ is norm continuous for each $a \in \mathcal{A}_n$ and $(id_t \otimes \alpha)(a)$ converges to a for $t \to \infty$. $\square$

Chapter 4

Arveson's Spectral C^*-Algebras

4.1 Product Systems

4.1.1 E_0-Semigroups and Product Systems

We review some results of [Ar 89],[Ar 90a] and [Ar 94b]. The motivation for Arveson's notion of product systems comes from the theory of E_0-semigroups.

Definition 4.1.1 *Let M be a von Neumann algebra acting on the separable Hilbert space H (in the sequel usually $\mathcal{B}(H)$). A family of normal $*$-endomorphisms $(\alpha_t)_{t\in\mathbb{R}_+}$ is called an e_0-semigroup (or just semigroup) on M if*

(i) $\alpha_t \circ \alpha_s = \alpha_{s+t}$, $\alpha_0 = id$ and α_t is not an automorphism for some $t > 0$.

(ii) For all $\xi, \eta \in H$ and $A \in M$ the map $t \mapsto \langle \xi, \alpha_t(A)\eta \rangle$ is continuous.

It is called an E_0-semigroup if

(iii) $\alpha_t(\mathbf{1}) = \mathbf{1}$ for each $t \in \mathbb{R}_+$.

For E_0-semigroups there are at least two important equivalence relations [1]:

Definition 4.1.2 *(i) E_0-semigroups α and β on M and N are called conjugate if there exists a $*$-isomorphism $\Theta : M \to N$ s.t. $\beta_t \circ \Theta = \Theta \circ \alpha_t$ for all $t > 0$.*

---
[1] E_0-semigroups have slightly better properties w.r.t cocycle conjugacy than e_0-semigroups

(ii) E_0-semigroups on M are called cocycle (or outer) conjugate if there exists a strongly continuous family $(U_t)_{t \in \mathbb{R}_+}$ of unitaries in the Hilbert space H s.t. $U_{s+t} = U_s \alpha_s(U_t)$ and $\beta_t(A) = U_t \alpha_t(A) U_t^*$ for each $A \in M$ and $s, t > 0$.

As mentionend before the corresponding cohomology group (cocycle conjugacy mod conjugacy) is rather subtle (c.f. instance [Ar 95] for some discussion in another setting). The only natural operation of E_0-semigroups is the tensor product. Any sort of index theory should be additive w.r.t. taking tensor products. In the sequel we only consider the case when M is equal to $\mathcal{B}(H)$. There is a well known correspondence between (normal) *-endomorphisms and Hilbert spaces in $M = \mathcal{B}(H)$, i.e. subspaces $\mathcal{S} \subseteq \mathcal{B}(H)$ s.t. $S^*T \in \mathbb{C}\mathbf{1}$ for $S, T \in \mathcal{S}$ and $\mathcal{S}$ is a Hilbert space with the scalar product $\langle S, T \rangle \mathbf{1} := S^*T$. The norm and the strong topology are the same on $\mathcal{S}$ ($T_\lambda \to T$ strongly implies $\|(T_\lambda - T)\xi\|^2 = \|T_\lambda - T\|^2 \|\xi\|^2 \to 0$ for each $\xi \in H$ thus convergence in norm). An onb (V_n) of $\mathcal{S}$ is a maximal sequence of isometries in $\mathcal{S}$ with pairwise orthogonal ranges. $\alpha(A) := \sum_n V_n A V_n^*$ defines a normal *-endomorphism of $\mathcal{B}(H)$ in this case[2]. Conversely, if α is such an endomorphism, then $\mathcal{E}_\alpha := \{T \in \mathcal{B}(H) | \alpha(A) T = TA \ \forall A \in \mathcal{B}(H)\}$ is a Hilbert space in $\mathcal{B}(H)$ and $\alpha(A) = \sum V_n A V_n^*$ where (V_n) is an onb in $\mathcal{E}$ [Ar 89 Prop.2.1]. The endomorphism is unital iff $\mathcal{E}_\alpha$ has full support ($:\Leftrightarrow \sum_n V_n V_n^* = \mathbf{1}$ for some onb).

For two Hilbert spaces $\mathcal{S}$ and $\mathcal{T}$ in $\mathcal{B}(H)$ the closed span of the set $\mathcal{ST} = \{ST | S \in \mathcal{S}, T \in \mathcal{T}\}$ is again a Hilbert space s.t. $(V_n W_m)$ is an onb of $\mathcal{ST}$ if $(V_n) \subseteq \mathcal{S}$ and $(W_m) \subseteq \mathcal{T}$ are onbs. Thus $[\mathcal{ST}]$ is the Hilbert space tensor product. If α and β are the endomorphisms corresponding to $\mathcal{E}_\alpha$ and $\mathcal{E}_\beta$, then $\mathcal{E}_{\alpha \circ \beta} = [\mathcal{E}_\alpha \mathcal{E}_\beta]$.

Each E_0-semigroup defines a family $E_\alpha(t) := \mathcal{E}_{\alpha_t}$ of Hilbert spaces with full support in $\mathcal{B}(H)$, and because α was assumed not to be automorphic $E_\alpha(t) \cap E_\alpha(s) = \{0\}$ for $t \neq s$. The union $E_\alpha = \bigcup_{t>0} E_\alpha(t)$ is a Borel subset of $\mathcal{B}(H)$ fibered over $\mathbb{R}_+$ and is closed under the product in $\mathcal{B}(H)$. It may be viewed as a "multiplicatively graded" set over $\mathbb{R}_+$ or a continuous tensor algebra in the sense that $E_\alpha(t) \otimes E_\alpha(s) \cong [E_\alpha(t) E_\alpha(s)] = E_\alpha(s+t)$. Notice

---

[2]for M a factor such endomorphisms are sometimes called inner

that in the full Fock space $\mathcal{F}(H)$ the union of the tensor powers $\bigcup_n H^{\otimes n}$ form a multiplicative set over $\mathbb{N}$ the product being the tensor product (c.f. Introduction).

Moreover, it follows from [Ar 89 Lemma 2.3] that for a nontrivial E_0-semigroup the family $E_\alpha(t)$ is measurable and the map $p : E_\alpha \to (0,\infty)$ is a trivial Borel fibration (compare Lemma 2.1.4). E_α is called a concrete product system.

The general definition of product systems is as follows:

Let E be a standard Borel space together with $p : E \to (0,\infty)$ Borel measurable s.t. $E(t) := p^{-1}(t)$ is a (separable) Hilbert space for each $t \in (0,\infty)$ and there is an isomorphism of Borel fibrations i.e. a Borel isomorphism $\Theta : E \to (0,\infty) \times E(t_0)$ for some $t_0 > 0$ (and hence any $t > 0$) s.t. the diagram

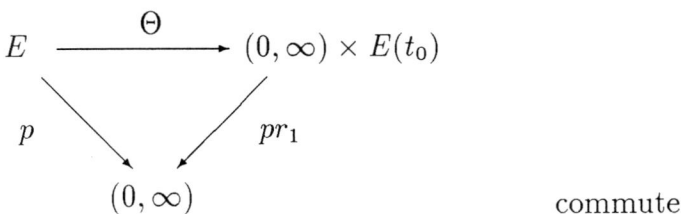

commutes.

Finally there is a jointly measurable associative multiplication $E \times E \to E$ satisfying

(i) $p(xy) = p(x) + p(y)$

(ii) $E(s+t)$ is spanned by $E(s)E(t)$ for every $s,t > 0$ and $\langle xy, x'y' \rangle = \langle x, x' \rangle \langle y, y' \rangle$ for any $x, x' \in E(s)$ and $y, y' \in E(t)$.

It follows that there is always a sequence e_n of Borel measurable sections in E s.t. $(e_n(t))_{n\in\mathbb{N}}$ is an onb of $E(t)$ for each $t > 0$. Such a sequence is called trivializing.

Definition 4.1.3 *E with the above properties is called a product system. A morphism of product systems $\Theta : E \to F$ is a fibre preserving Borel map s.t. $\Theta(xy) = \Theta(x)\Theta(y)$ and $\Theta|E(t) : E(t) \to F(t)$ is a bounded linear map. A representation of a product system is a Borel measurable map $\phi : E \to \mathcal{B}(H)$*

($\mathcal{B}(H)$ equiped with its strong Borel structure) s.t. $\phi(xy) = \phi(x)\phi(y)$ for any $x, y \in E$ and $\phi(x)^*\phi(x') = \langle x, x' \rangle \mathbf{1}$ for $x, x' \in E(t)$.

For instance the standard Borel space $(0, \infty) \times \mathbb{C}$ with $p = pr_1$ and the product $(s, \lambda)(t, \mu) = (s + t, \lambda\mu)$ is a product system. It is called the trivial product system and occurs when looking at the trivial e_0-semigroup $\alpha(A) = V_t A V_t^*$ where V_t is a strongly continuous nonunitary semigroup of isometries.

There is a natural tensor product of product systems $E \otimes F$ defined by $(E \otimes F)(t) := E(t) \otimes F(t)$.

For any product system E there is the opposite product system which is the same Borel fibration $p : E \to (0, \infty)$ with reversed multiplication.

Example 4.1.4 The measurable sections ξ to $p : E \to (0, \infty)$ s.t. $\int_0^\infty dt \, \|\xi(t)\|^2 < \infty$ may be taken as an admissible set defining a direct integral (c.f. Appendix B) $L^2(E) := \int^\oplus E(t) dt$ on which E acts by

$$l(v)(\xi)(x) = \begin{cases} v\xi(x - t) & \text{if } x > t \\ 0 & \text{otherwise} \end{cases}$$

The admissible set is invariant under this action and this defines the so called regular representation l of E.

Summarizing our discussion above we have (c.f.[Ar 89]):

Proposition 4.1.5 For any E_0-semigroup α the set $E_\alpha = \bigcup_{t>0} \{t\} \times E_\alpha(t) \subseteq (0, \infty) \times \mathcal{B}(H)$ with the Borel structure coming from the strong topology is a product system.

Thus each concrete product system is an abstract one.

The role of a product system is that of a spectrum of an e_0-semigroup:

Proposition 4.1.6 For a product system E and a representation $\phi : E \to \mathcal{B}(H)$, H separable, there is a unique e_0-semigroup α s.t.

$$\alpha_t(A)\phi(v) = \phi(v)A$$

for all $A \in \mathcal{B}(H)$ and ϕ is an isomorphism between E and E_α in the sense that $E_\alpha(t) = \phi(E(t))$.

Proof: [Ar 89 Prop.2.7] □

The E_0- and e_0-semigroups now correspond to essential and nonessential representations of product systems: A representation $\phi : E \to \mathcal{B}(H)$ is called essential [Ar 90c] if $\overline{\phi(E(t))H} = H$ for each $t > 0$ and singular if $\bigcap_{t\in\mathbb{R}_+} \phi(E(t))H = 0$. For instance the regular representation $l : E \to L^2(E)$ from Ex.4.1.4 is singular and it is the only obvious representation of a product system in general. The problem solved in [Ar 90c] is to show that essential representations of product systems always exist. Thus it is no restriction to consider E_0-semigroups. Therefore each abstract product system is a concrete one.

It turns out that isomorphy of product systems corresponds to cocycle conjugacy of E_0-semigroups:

Proposition 4.1.7 *Let α, β be E_0-semigroups on $\mathcal{B}(H)$ and $\mathcal{B}(K)$ respectively.*

(i) *If α and β are conjugate, then E_α and E_β are isomorphic and the representations of the corresponding product systems are unitarily equivalent.*

(ii) *If $H = K$, then α and β are cocycle conjugate iff E_α and E_β are isomorphic.*

Proof: (c.f. [Ar 89 sec.2 and 3.18])
(i): Θ is unitarily implemented and the multiplication by the unitary maps E_α to E_β.
(ii): If (U_t) is a cocycle for α and β, then $U_t E_\alpha(t) = E_\beta(t)$. Conversely, if $\Theta : E_\alpha \to E_\beta$ is an isomorphism and $(V_n(t))$ a trivialization of E_α (i.e. $t \mapsto V_n(t)$ are measurable families of isometries s.t. $(V_n(t))$ are onb's of $E_\alpha(t)$), then, because α and β are unital, $U_t := \sum_{n=0}^{\infty} \Theta(V_n(t))V_n(t)^*$ is a unitary cocycle. □

A unit in a product system is a nonzero section $t \mapsto u(t) \in E(t)$ s.t. $u(t+s) = u(t)u(s)$ $\forall s, t > 0$. For units u and v the function $f : t \mapsto \langle u(t), v(t) \rangle$ fulfils $f(s+t) = f(s)f(t)$ and thus is of the form $e^{\gamma(u,v)t}$. If $\mathcal{U}_E$ is the set of units of E, then the index of E is defined as the dimension minus 1 of the Hilbert space given by the kernel $\langle u(t), v(t) \rangle$ on $\mathcal{U}_E$ for any $t > 0$. In a representation ϕ of E with α the corresponding e_0-semigroup a normalized unit ($\|u(t)\| = 1 \ \forall t > 0$) becomes a strongly continuous semigroup of isometries $U(t) = \phi(u(t))$ s.t. $\alpha_t(A)U(t) = U(t)A \ \forall t > 0, \ A \in \mathcal{B}(H)$. In the sequel we mean by unit usually a normalized unit.

A. The CCR-Flows

We illustrate the foregoing in the easiest case:

Define the exponential product systems E_n as follows: $E_n(t) := (t, \mathcal{F}^s(L^2((0,t), \mathbb{C}^n)))$, $E_n := \{(t, \xi) | t \in (0, \infty), \xi \in \mathcal{F}^s(L^2((0,t), \mathbb{C}^n))\} \subseteq \mathbb{R}_+ \times \mathcal{F}^s(L^2(\mathbb{R}_+, \mathbb{C}^n))$ considered as a Borel subspace.

Using the unitary $U(t, t_0) := exp(S(t, t_0)) : E(t) \to E(t_0)$ where $S(t, t_0) : L^2((0, t_0), \mathbb{C}^n) \to L^2((0, t), \mathbb{C}^n)$ is defined by $f \mapsto \sqrt{\frac{t_0}{t}} f(\frac{t}{t_0} \cdot)$, we obtain an isomorphism $E_n \to (0, \infty) \times E(t_0)$ between Borel fibrations.

As product in E_n we define

$$(t, exp(f))(s, exp(g)) := (s + t, exp(f + S_t g)) = (s + t, exp(f) \otimes exp(S_t g))$$

where $U(t) = exp(S_t)$ is considered as unitary between $\mathcal{F}^s(L^2((0, s), \mathbb{C}^n))$ and $\mathcal{F}^s(L^2((t, s+t), \mathbb{C}^n))$. This definition makes sense for any $(t, \xi) \in E_n(t)$ and $(s, \eta) \in E_n(s)$. The product is measurable (even continuous) and we obtain a product system for each $n \in \mathbb{N}$. These product systems E_n are called the exponential product systems. They are also the only known explicit examples of product systems. Note that $E_n = E_1 \otimes \ldots \otimes E_1$ for the tensor product of product systems. In [Ar 89 sec.6] it is shown that E_n has index n and the E_n are exactly the product systems generated by their units in the sense that $E_n(t) = [\{u_1(t_1) \ldots u_k(t_k) | t_1 + \ldots + t_k = t, \ u_1, \ldots, u_k \in \mathcal{U}_E\}]$ for each $t > 0$.

We have the representation of E_n on $\mathcal{F}^s(L^2(\mathbb{R}_+, \mathbb{C}^n))$ given by

$$\phi(e(t))(\eta) := e(t) \otimes U(t)\eta$$

where $U(t)$ is considered as a unitary beween $\mathcal{F}^s(L^2(\mathbb{R}_+, \mathbb{C}^n))$ and $\mathcal{F}^s(L^2((t, \infty), \mathbb{C}^n))$. Now define an E_0-semigroup α on $H := \mathcal{F}^s(L^2(\mathbb{R}_+, \mathbb{C}^n))$

as follows:
$$\alpha_t(A)(e(t) \otimes U(t)\eta) = e(t) \otimes U(t)(A\eta)$$
if $e(t) \in E_n(t)$, $\eta \in \mathcal{F}^s(L^2(\mathbb{R}_+, \mathbb{C}^n))$ and $A \in \mathcal{B}(H)$. Then $E_\alpha(t)$ consists exactly of those $T \in \mathcal{B}(H)$ of the form
$$T_e \eta = e \otimes U(t)\eta$$
with $e \in E_n(t)$:

$T_e \in E_\alpha(t)$ for each $e \in E_n(t)$ because $\alpha_t(A)T_e\eta = \alpha_t(A)(e \otimes U(t)\eta) = e \otimes U(t)A\eta = T_e A\eta$.

If, on the other hand, e_k is a trivializing sequence in E_n, then $T_{e_k(t)} =: V_k(t)$ is a family of isometries with orthogonal ranges s.t.

$$\sum_k V_k(t)V_k(t)^* = \sum_k P_{e_k(t) \otimes \mathcal{F}^s(L^2((t,\infty),\mathbb{C}^n))} = \mathbf{1}$$

hence is an onb of $E_\alpha(t)$. Therefore $e \mapsto T_e$ is one to one between $E_n(t)$ and $E_\alpha(t)$. Now α and ϕ are exactly in the correspondence of Prop.4.1.6.

Note that the action of α on certain finite rank operators is given by

$$\alpha_t(\Theta_{exp(f),exp(g)}) = \mathbf{1}_{(0,t)} \otimes \Theta_{exp(S_t f),exp(S_t g)}$$

which could also be used to define α perhaps most suggestively. It turns out that this E_0-semigroup has the numerical index n. The semigroup corresponding to $n = 1$ is Powers' CAR-flow because one can map the antisymmetric Fock space to the symmetric Fock space over $L^2(0,t)$ by noticing that antisymmetric and symmetric L^2-functions on $(0,t)^k$ are both given by their values on $\{(t_1, \ldots, t_k) | 0 < t_1 < \ldots < t_k < t\}$ [PoRo 89].

B. Decomposable Product systems

There is a close connection between product systems and continuous tensor decompositions considered before.

Proposition 4.1.8 *Let E be a product system and fix $t > 0$. Then $(E(t) \cdot \mathcal{B}_0(0,t))$ is a continuous tensor decomposition if we define the local*

Hilbert spaces by $\mathcal{H}_{[x,y]} = E(y-x)$ *for* $[x,y] \subseteq (0,t)$ *and* $\mathcal{H}_{(0,y)} = E(y)$ *for* $y \in (0,t)$. *If* $P = \{I_i\} \leq Q = \{J_k\}$ *are partitions into subintervals, define* $\varphi_P(\otimes_i x_i) := \prod_i x_i$ *and* $\varphi_{PQ}(\otimes_i x_i) := \otimes_k (\prod_{I_i \subseteq J_k} x_i)$. *Then this defines a continuous tensor decomposition. If* E *and* F *are isomorphic product systems, then* $(E(t), \mathcal{B}_0(0,t))$ *and* $(F(t), \mathcal{B}_0(0,t))$ *are isomorphic tensor decompositions for all* $t > 0$.

Proof: It is clear that we obtain a tensor decomposition. The continuity of $((\mathcal{B}(\mathcal{H}_t) \otimes \mathbf{1})_t)$ follows from [Ar 89 Lemma 2.3] and [Ar 90a Prop. 5.5]. The condition for weak continuity may be seen as follows: By [Ar 90c Cor.5.17], we may assume $E = E_\alpha$, α an E_0-semigroup on $\mathcal{B}(H)$. By [Ar 89 Lemma 2.3] we can find a trivialization $(V_n(t)) \subseteq \mathcal{B}(H)$ (i.e. a family of isometries) s.t. $(0, \infty) \ni t \mapsto V_n(t)$ is strongly continuous. Then $(V_n(s)V_m(t-s))$ $(s \in (0,t))$ is an onb for $E_\alpha(t)$ as required. Finally suppose E and F are isomorphic via Θ. Let $t > 0$ and define the local maps $\varphi_{[x,y]} = \Theta | E(y-x)$. Then $\varphi_{(0,t)}$ decomposes into tensor products of the local maps. $\square$

Definition 4.1.9 *Let* E *be a product system.* $\zeta \in E(t)$ *for* $t > 0$ *is called decomposable if for every* $s \in (0,t)$ *there are* $\xi \in E(s)$, $\eta \in E(t-s)$ *s.t.* $\zeta = \xi\eta$. *If* $D(t)$ *denotes the set of decomposable vectors in* $E(t)$, *then* E *is called decomposable if* $[D(t)] = E(t)$ *for some (equivalently each)* $t > 0$.

Thus we have a new proof of [Ar 94b Thm.11.1]:

Theorem 4.1.10 *Each decomposable product system is exponential.*

Proof: Let E be decomposable. By Thm.2.2.4, we have $(E(t), \mathcal{B}_0(0,t))$ is exponential i.e. $E(t) = \otimes_{(0,t)}(\mathcal{K}_x, \mu)$ for some continuous measure μ on $(0,t)$ and measurable family $(\mathcal{K}_x)_{x \in (0,t)}$. Moreover, $E(s) = \otimes_{(0,s)}(\mathcal{K}_x, \mu|(0,s))$ for $s < t$ by the definition of the multiplication. We have $E(t) = E(s)\hat{\otimes}E(t-s)$ as continuous tensor decompositions where $\hat{\otimes}$ is the tensor product of tensor decompositions mentioned in Rem.2.1.2.(v). Both sides are exponential. If $[\mu_n]$ are the measure classes of $\int_{(0,t)}^{\oplus} \mathcal{K}_x d\mu(x)$ for the multiplicity n, put $\mu_{n,s} := \mu_n|(0,s)$. Let T_s be the the translation on $\mathbb{R}_+$ with $s > 0$. Then according to 2.2.4 we must have

$$[\mu_n] = [\mu_{n,s} + T_s\mu_{n,t-s}]$$

By Appendix Rem.5.(ii), $[\mu_n] = 0$ except for at most one $n \in \mathbb{N} \cup \{\infty\}$ and the one nonzero $[\mu_n]$ is the Lebesgue measure class. But that means $E = E_n$.
□

We remark that the rearranged exponential tensor decomposition in sec.2.1.4 suggests the construction of E_0-semigroups of type III as an inductive limit of a sequence of type I E_0-semigroups $\alpha^{(l)}$ on type I-subfactors $M_l = \mathcal{B}(H_l)$ and $\alpha^{(l+1)}$ extends $\alpha^{(l)}$. In fact it is easy to define a unital semigroup α_t on a weakly dense subalgebra of some $\mathcal{B}(H)$. However, α_t fails to be normal.

4.2 The Spectral C^*-Algebra $C^*(E)$ of a Product System

4.2.1 The Wiener Hopf C^*-Algebra

The results in this section are more or less folklore facts. Prop.4.2.1 is sometimes proved by using Naimark's dilation theorem (c.f.[Fil 68]). All other results with the exception of Cor.4.2.7, Prop.4.2.8 (due to Arveson) and the appearently unnoticed Cor.4.2.10 seem to be well known (c.f. [BöSi 89 Ch.9] for a summary of some results).

Proposition 4.2.1 (*v.Neumann-Wold-decomposition*). *Let $(u_t)_{t \in \mathbb{R}_+}$ be a strongly continuous semigroup of isometries on a Hilbert space H. Then there is a unique decomposition*

$$H = H_1 \oplus H_2 \text{ and } u_t = v_t \oplus w_t$$

where v_t is a strongly continuous unitary semigroup on H_1 and w_t is a multiple of the onesided shift T_t on $L^2(\mathbb{R}_+)$.

<u>Proof:</u> Let $p_t = u_t u_t^*$. Then p_t is monotone decreasing and strongly continuous. Let $p := s - \lim_{t \to \infty} p_t$ and $p' := 1 - p$, $H_1 := pH$, $H_2 := p'H$. For $\xi \in H$ we have $u_s p \xi = \lim_{t \to \infty} u_{s+t} u_t^* \xi = \lim_{r \to \infty} u_r u_r^* u_s \xi$. Hence $u_s p = p u_s$ for each $s \in \mathbb{R}_+$. Put $v_t := p u_t p$, $w_t = p' u_t p'$. Then $u_t = v_t \oplus w_t$, and v_t,

w_t are isometric, strongly continuous semigroups on H_1 and H_2 respectively. $u_t u_t^* p = p$ implies v_t unitary and $w_t w_t^* \leq p' u_t u_t^* p'$ implies $s-\lim_{t \to \infty} w_t w_t^* = 0$. Now $P_t := p_t - p$ is a decreasing family of projections s.t. $P_0 = p'$, $P_\infty = 0$. Hence P_t defines a spectral resolution in $H_2 = p'H$ (c.f. Appendix Def.11). By Appendix Thm.13, P_t defines a direct integral $\int_{\mathbb{R}_+}^\oplus \mathcal{K}_t d\mu(t)$ s.t. $P_a - P_b = m(\chi_{[a,b)})$ with m the diagonal multiplication operator. The direct integral is specified by the disjoint measure classes $[\mu_1], \ldots, [\mu_\infty]$. Because $m(\chi_{[a+t,b+t)}) = w_t m(\chi_{[a,b)}) w_t^*$ by the definition of P_t the measures μ_i and $\mu_i(\cdot + t)$ are equivalent for each $t > 0$. By Appendix Rem.5.(i), there is only one such measure class on $\mathbb{R}_+$ namely the Lebesgue measure class. Hence all but one measure class $[\mu_{i_0}]$ must be zero and $[\mu_{i_0}] = [dt]$. This implies $w_t \cong T_t^{\oplus i_0}$ where $\cong$ means unitary equivalence. □

Corollary 4.2.2 *For any two nonunitary semigroups of isometries u_t, s_t the map sending $\int dt\, f(t) u_t$ to $\int dt\, f(t) s_t$, $f \in L^1(\mathbb{R}_+)$ extends to an isomorphism of the generated C^*-algebras.*

<u>Proof:</u> It suffices to show that if V_t is any strongly continuous unitary group, then $\|\int dt\, f(t) V_t\| \leq \|\int dt\, f(t) T_t\|$ and $\|\int\int dx dy\, F(x,y) V_x V_y^*\| \leq \|\int\int dx dy\, F(x,y) T_x T_y^*\|$ for each $f \in L^1(\mathbb{R}_+)$, $F \in L^1(\mathbb{R}_+^2)$ and T_t the onesided shift. If $V_x = \int e^{ixt} dE(x)$ and $F(x,y) = \sum_{i,j}^k f_i(x) g_j(-y)$, then $\int\int dx dy\, F(x,y) V_x V_y^* = 2\pi \sum \int \hat{f}_i(t) \hat{g}_j(t) dE(t)$, $\int dx\, f(x) V_x = \sqrt{2\pi} \int \hat{f}(t) dE(t)$. Hence we may assume that V_t is the two sided shift S_t on $L^2(\mathbb{R})$. The assertion follows from the exact sequence of the Wiener-Hopf-algebra using that the symbol map $\pi : \mathcal{W} \to C_0(\mathbb{R})$ is given by the Fourier transform. (c.f. Prop.4.2.8) □

Lemma 4.2.3 *The set $\text{span}\{x \mapsto x^n e^{-x} | n \in \mathbb{N}\}$ is dense in $L^p(\mathbb{R}_+)$ for $p = 1, 2$.*

<u>Proof:</u> For $L^2(\mathbb{R}_+)$ this is very classical [CoHi p.81-82]. For $L^1(\mathbb{R}_+)$ let $f \in L^\infty(\mathbb{R}_+)$ s.t. $\int dt\, t^n f(t) e^{-t} = 0$ for each $n \in \mathbb{N}$. The function $z \mapsto F(z) = \int_0^\infty dt\, f(t) e^{izt}$ is analytic in the open upper half plane and $i^n \int dt\, t^n f(t) e^{-t}$ is the value of its n-th complex derivative at $z = i$. Hence F is identically 0. For each $0 < \alpha < 1$ consider the $L^1(\mathbb{R}_+)$-function $f_\alpha(t) := e^{-\alpha t} f(t)$. Then $\sqrt{2\pi} \hat{f}_\alpha(t) = F(t + i\alpha)$ and therefore $\hat{f}_\alpha = 0$ which implies that $f_\alpha(t) = f(t) e^{-\alpha t} = 0$ hence $f(t) = 0$ for almost all $t \in \mathbb{R}_+$. □

ARVESON'S SPECTRAL C*-ALGEBRAS

The orthogonalization of $(x \mapsto x^n e^{-x})$ yields the following. Let $L_n(x) = \sum_{k=0}^n (-1)^k \binom{n}{k} n(n-1)\ldots(k+1) x^k$ be the n-th Laguerre polynomial ([CoHi p.79]) and put $l_n(x) = \frac{L_n(x)}{n!} = \sum_{k=0}^n \frac{(-1)^k}{k!} \binom{n}{k} x^k$. Then $x \mapsto \varphi_n(x) := e^{-\frac{x}{2}} l_n(x)$ is an onb of $L^2(\mathbb{R}_+)$ as well as $x \mapsto \frac{1}{\sqrt{2}} \varphi_n(2x)$.

Define $V = 1 - 2 \int_0^\infty dt\, e^{-t} T_t$ where T_t is the onesided shift on $L^2(\mathbb{R}_+)$. With t the onesided shift on $l^2(\mathbb{N})$ we have :

Proposition 4.2.4 V is an isometry and the C^*-algebras $C^*(t)$ and $C^*(\{\int dt\, f(t) T_t | f \in L^1(\mathbb{R}_+)\})^\sim$ are unitarily equivalent.

<u>Proof:</u> $V(x \mapsto x^n e^{-x}) = x \mapsto (x^n - \frac{2}{n+1} x^{n+1}) e^{-x}$. Hence

$$
\begin{aligned}
(V\varphi_n)(2x) &= \sum_{k=0}^n \frac{(-2)^k}{k!} \binom{n}{k} (x^k - \frac{2}{k+1} x^{k+1}) e^{-x} \\
&= e^{-x} + \frac{(-2)^{n+1}}{(n+1)!} x^{n+1} e^{-x} \\
&\quad + \sum_{k=1}^n \left[\frac{(-2)^k}{k!} \binom{n}{k} - \frac{(-1)^{k+1} 2^k}{(k-1)!} \binom{n}{k-1} \frac{1}{k} \right] x^k e^{-x} \\
&= \sum_{k=0}^{n+1} \frac{(-2x)^k}{k!} \binom{n+1}{k} e^{-x} \\
&= \varphi_{n+1}(2x)
\end{aligned}
$$

Thus V is the onesided shift. For surjectivity note

$$
\left(\int_0^\infty dt\, e^{-t} T_t \right)^k = \int_0^\infty ds \int_0^s dt_{k-1} \int_0^{t_{k-1}} dt_{k-2} \ldots \int_0^{t_2} dt_1\, e^{-s} T_s
$$
$$
= \frac{1}{(k-1)!} \int_0^\infty ds\, s^{k-1} e^{-s} T_s
$$

because the volume of the simplex with k vertices is $\frac{1}{(k-1)!}$. By Lemma 4.2.3 we can approximate each symbol function in L^1. $\square$

Corollary 4.2.5 *For s a nonunitary isometry and u_t a nonunitary strongly continuous semigroup of isometries, $C^*(s)$ and $C^*(\{\int dt\, f(t) u_t \mid f \in L^1(\mathbb{R}_+)\})^\sim$ are isomorphic.*

Remark 4.2.6 $C^*(t-1)$ *contains the finite rank operators.*

Proof: $e_{ij} = p_i t^{i-j} p_j$ ($t^{-1} := t^*$) and $p_k = t^k t^{*k} - t^{k+1} t^{*k+1} = (t-1+1)^k (t-1+1)^{*k} - (t-1+1)^{k+1}(t-1+1)^{*k+1}$ all lie in $C^*(t-1)$ and are matrix units for the onb in $l^2(\mathbb{N})$. □

Corollary 4.2.7 *The C^*-algebra*

$$\mathcal{W}_0 := C^*\left(\left\{\int\int dxdy\, f(x,y)T_x T_y^* | f \in L^1(\mathbb{R}_+^2)\right\}\right)$$

contains the compacts.

Proof: First remark that because T_x is strongly continuous $v_\varepsilon \xi = \frac{1}{\varepsilon}\int_0^\varepsilon dx\, T_x \xi \xrightarrow{\varepsilon \to 0} \xi$ if $\xi \in L^2(\mathbb{R}_+)$. The same holds for v_ε^*. Now let $F = P(t-1, t^*-1)$, P a polynomial, $t = 1 - 2\int dx\, e^{-x} T_x$ s.t. F is a finite rank matrix in the onb given by $x \mapsto 2^{-1/2}\varphi_n(2x)$ as above. Then $v_\varepsilon F v_\varepsilon^* \in \mathcal{W}_0$ for each $\varepsilon > 0$ and converges to F in norm. □

The following will be useful later on and has been mentioned in [Ar 90a] (without proof).

Proposition 4.2.8 *For any nonunitary strongly continuous semigroup u_t of isometries the C^*-algebras $\mathcal{W} := C^*(\{\int dt\, f(t)u_t | f \in L^1(\mathbb{R}_+)\})$ and $\mathcal{W}_0 := C^*(\{\int dxdy\, f(x,y)u_x u_y^* | f \in L^1(\mathbb{R}_+^2)\})$ are the same.*

Proof: By Cor.4.2.2, we may assume $u_t = T_t$. $\mathcal{W}_0 \subseteq \mathcal{W}$ contains the compacts $\mathcal{K}$. Thus $\mathcal{W}_0/\mathcal{K} \subseteq \mathcal{W}/\mathcal{K}$ and the latter is known to be isomorphic to $C_0(\mathbb{R})$. $\pi: \mathcal{W} \to C_0(\mathbb{R})$ is given by the Fourier transform:

$$\pi\left(\int\int dxdy\, f(x,y)T_x T_y^*\right) = t \mapsto \int_0^\infty \int_0^\infty dxdy\, f(x,y)e^{i(x-y)t}$$

$$= \int_0^\infty dx \int_{-\infty}^x dz\, f(x, x-z)e^{izt}$$

If we take $h \in L^1(\mathbb{R})$ s.t. $supp(h) \subseteq [-t_0, t_0]$ and $g = \chi_{[t_0, t_0+1)}$, we can find $f \in L^1(\mathbb{R}_+^2)$ s.t. $f(x, x-z) = g(x)h(z)$. Then

$$\int_0^\infty dx \int_{-\infty}^x dz\, f(x, x-z)e^{izt} = \hat{h}(t)$$

ARVESON'S SPECTRAL C*-ALGEBRAS

Thus $\pi(\mathcal{W}_0) = C_0(\mathbb{R})$. There is the following commutative diagram with vertical maps being inclusions

$$\begin{array}{ccccccccc} 0 & \longrightarrow & \mathcal{K} & \longrightarrow & \mathcal{W} & \longrightarrow & C_0(\mathbb{R}) & \longrightarrow & 0 \\ & & \uparrow & & \uparrow & & \uparrow & & \\ 0 & \longrightarrow & \mathcal{K} & \longrightarrow & \mathcal{W}_0 & \longrightarrow & C_0(\mathbb{R}) & \longrightarrow & 0 \end{array}$$

The inclusion in the middle must therefore be an equality. □

Lemma 4.2.9 *For each $f \in L^1(\mathbb{R}_+^2)$ and $t \in \mathbb{R}_+$ the operator*

$$\int_0^\infty dx \int_0^\infty dy \, f(x,y) \, T_x P_t T_y^*$$

is compact, where P_t is the projection defined by $P_t = \mathbf{1} - T_t T_t^*$.

Proof:

(i) Let $f \in C_c(\mathbb{R}_+)$. Then $A = \int dx \, f(x) \, P_t T_x P_t$ is the integral operator with kernel

$$k(s,r) = \begin{cases} f(s-r) & \text{if } r < s < t \\ 0 & \text{otherwise} \end{cases}$$

Similarly, also $\int dx \, f(x) \, P_s T_x P_t$ is compact for $s, t > 0$.

(ii) $\int dx \, f(x) \, T_x P_t = \int dx \, f(x) \, P_{t+T} T_x P_t$ if $\mathrm{supp} f \subseteq [0,T] \subseteq \mathbb{R}_+$ in particular also this integral is compact.

(iii) Taking products like $\left(\int dx \, f(x) P_{T+t} T_x P_t\right)\left(\int dy \, g(y) P_t T_y^* P_{T+t}\right)$ and approximating we obtain: For $F \in C_c(\mathbb{R}_+^2)$, $\mathrm{supp} F \subseteq [0,T] \times [0,T]$ the operator $\int dx \int dy \, F(x,y) P_{t+T} T_x P_t T_y^* P_{t+T} = \int dx \int dy \, F(x,y) T_x P_t T_y^*$ is compact. Now approximate L^1-functions by C_c-functions in the L^1-norm. □

Corollary 4.2.10 u_t *is a two sided multiplier of $\mathcal{W}_0 = \mathcal{W}$ for any $t > 0$.*

Proof: (There is an alternative argument as in Prop.4.2.20 using Prop.4.2.8)
We may assume $u_t = T_t$ and show that right multiplication by T_t maps double integrals into elements in $\mathcal{W}_0$. For each $t > 0$, let $P^t = T_t T_t^* = 1 - P_t$ and $F \in L^1(\mathbb{R}_+^2)$. Then

$$\int dx \int dy F(x,y) T_x T_y^* = \int dx \int dy [F(x,y) T_x P_t T_y^* + F(x,y) T_x P^t T_y^*]$$

The first summand gives a compact operator by 4.2.9. For the second we get after multiplication by T_t from the right

$$\int dx \int dy\, F(x,y) T_x P^t T_y^* T_t = \int dx \int dy\, F(x,y) T_{x+t} T_y^*$$

which lies in $\mathcal{W}_0$. Because $\mathcal{K} T_t \subseteq \mathcal{K}$ and $\mathcal{K} \subseteq \mathcal{W}_0$ we infer that T_t is a right multiplier. Of course, it is also a left multiplier. □

4.2.2 The Involutive Banach Algebra $L^1(K_E)$

In this section we recall the definition of the involutive Banach algebra $L^1(K_E)$ from [Ar 90a]. We then show the existence of approximate units in these algebras which will be applied in sec.4.3.2. for inducing representations. Furthermore, we show the existence of approximate units in some of the nonselfadjoint algebras $L^1(E)$ which will be applied in Prop.4.2.19.

Let E be a product system. Consider families of compact operators $(K(x,y))_{x,y \in \mathbb{R}_+}$, $K(x,y) \in \mathcal{K}(E(y), E(x))$. According to the definition of a product system we have an isomorphism $\Theta : E \to E(t_0) \times (0,\infty)$ of Borel fibrations of Hilbert spaces s.t.

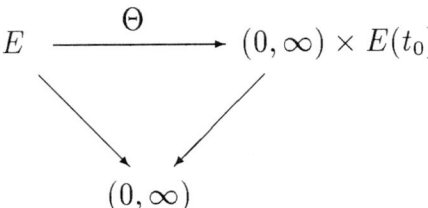

commutes. $U_x := \Theta|E(x) : E(x) \to E(t_0)$ is unitary. If (e_n) is trivializing for E, then U_x is defined by sending $e_n(x)$ to $e_n(t_0)$. The family $(K(x,y))_{x,y \in (0,\infty)}$ is called measurable if

$$(x,y) \mapsto U_x K(x,y) U_y^{-1} : E(t_0) \to E(t_0)$$

ARVESON'S SPECTRAL C^*-ALGEBRAS

is measurable as a family of compact operators.

The Banach algebra $L^1(K_E)$ is the set of measurable families $(K(x,y))_{x,y\in\mathbb{R}_+}$ as above s.t. $\int dx \int dy \, \|K(x,y)\| < \infty$. Thus as a Banach space $L^1(K_E)$ may be identified with $L^1(\mathbb{R}_+ \times \mathbb{R}_+, \mathcal{K}(E(t_0)))$ via

$$((x,y) \mapsto K(x,y)) \longleftrightarrow ((x,y) \mapsto U_x K(x,y) U_y^*)$$

equiped with the following product:

For $x, y, \lambda \in (0, \infty)$ we have

$$K(x,y) \otimes \mathbf{1}_\lambda : E(y) \otimes E(\lambda) = E(y+\lambda) \to E(x) \otimes E(\lambda) = E(x+\lambda)$$

Define for $K, G \in L^1(K_E)$:

$$(KG)(x,y) = \int_0^\infty ds \left[\int_0^{min(s,y)} d\lambda \; K(x,s)(G(s-\lambda, y-\lambda) \otimes \mathbf{1}_\lambda) + \right.$$

$$\left. + \int_0^{min(x,s)} d\lambda \; (K(x-\lambda, s-\lambda) \otimes \mathbf{1}_\lambda) G(s,y) \right]$$

The involution of $K \in L^1(K_E)$ is given by $(K^*)(x,y) := K(y,x)^*$. It can be checked that this product turns $L^1(K_E)$ into an involutive Banach algebra. [Ar 90a p.149]

Definition 4.2.11 *The spectral C^*-algebra of a product system $C^*(E)$ is the enveloping C^*-algebra of $L^1(K_E)$.*

Thus $C^*(E)$ is a separable C^*-algebra the representations of which may be identified with the contractive representations of $L^1(K_E)$.

We adopt Arveson's notation: for L^1-sections $f, g \in L^1(E)$ in E there is a rank 1 section in $L^1(K_E)$ defined by

$$(f \otimes \overline{g})(x,y) := \Theta_{f(x), g(y)}$$

The span of them are called the finite rank elements.

Proposition 4.2.12 (i) ([Ar 90a Prop.2.13]) *The finite rank elements form a dense selfadjoint subset of $L^1(K_E)$.*

(ii) The subset of sections in $L^1(K_E)$ which are finite dimensional a.e. form a dense subalgebra.

Proof: (i): The identification of $L^1(K_E)$ and $L^1(\mathbb{R}_+^2, \mathcal{K})$ as Banach spaces shows the density of the finite rank elements (compare [Ar 90a Prop.2.13]).

(ii): One only has to see that it is a subalgebra. This follows by looking at the formula for the product in $L^1(K_E)$. □

Let $U(x,y) := U_x^{-1} U_y : E(y) \to E(x)$. Then $U(x,y) e_n(y) = e_n(x)$ for $n \in \mathbb{N}$ and each trivialization defines a family of such operators. We may define $U(x,y) \otimes \mathbf{1}_\lambda : E(y+\lambda) \to E(x+\lambda)$ by the requirement $uv \mapsto (U(x,y)u)v$ where $u \in E(y)$, $v \in E(\lambda)$. $\mathbf{1}_\lambda \otimes U(x,y)$ is defined similarly by $vu \mapsto v(U(x,y)u)$. [3]

Lemma 4.2.13 *(i) For any given family U_x as above $P_N(x,y) := \sum_{n=0}^N e_n(x) \otimes \overline{e_n(y)} \xrightarrow{N \to \infty} U(x,y)$ weakly.*

(ii) There is a dense subset $\mathcal{D}$ in $L^1(K_E)$ s.t. for any $F \in \mathcal{D}$ there is $c > 0$ with supp $F \subseteq (c, \infty) \times (c, \infty)$ and a choice of a family $U_x : E(x) \to E(t_0)$ s.t. the section F^ε defined by

$$F^\varepsilon(x,y) := (U(x-\lambda, x+\varepsilon-\lambda) \otimes \mathbf{1}_\lambda) F(x+\varepsilon, y)$$

converges for each $0 < \lambda < c$ pointwise in norm and, as a section in L^1-norm, to F if $\varepsilon \to 0$. It varies continuously in λ pointwise and as a section in the L^1-sense.

(iii) If U_x is as in (ii) and $F \in \mathcal{D}$, let $c > 0$ be as in (ii). For $0 < \varepsilon < c$ define the sections

$$F^{N,\varepsilon}(x,y) = \tfrac{1}{\varepsilon^2} \int_x^{x+\varepsilon} ds \int_{max(0,x-\varepsilon)}^x d\lambda \, (P_N(x-\lambda, s-\lambda) \otimes \mathbf{1}_\lambda) F(s,y)$$

where $P_N(x,y)$ is defined w.r.t. $U(x,y)$. Then there is a sequence of natural numbers N_l and a zero sequence of positive numbers ε_l s.t. F^{N_l, ε_l} converges to F for $l \to \infty$ in L^1-norm with Lebesgue integrable majorant.

---

[3] For the following it would be more appropriate to redefine product systems slightly considering them as topological rather than Borel spaces.

Proof: (i): Clear.

(ii): First we may replace E by its opposite product system (i.e. E with the reverse product c.f. remark after Def.4.1.3). Then we have to consider $F^\varepsilon(x,y) = (\mathbf{1}_\lambda \otimes U(x-\lambda, x+\varepsilon-\lambda))F(x+\varepsilon,y)$. $\mathcal{D}$ is now defined as a set of in some sense continuous sections and we require a bit more structure for that. Let $E = E_\alpha$ for some E_0-semigroup α on $\mathcal{B}(H)$ (without loss by [Ar 90c Cor.5.17]). Let U_x be as in [Ar 89 Lemma 2.3] which is a strongly continuous family of unitaries in $\mathcal{B}(H)$ and multiplication by U_x^* maps $E_\alpha(x)$ to $E_\alpha(t_0)$[4]. Thus $U(x,y) = U_x U_y^*$ is a strongly continuous two parameter family in $\mathcal{B}(H)$ s.t. $U(x,y) \to \mathbf{1}$ if $x \to y$ and $\alpha_x(A) = U(x,y)\alpha_y(A)U(y,x)$ for each $A \in \mathcal{B}(H)$ (compare Lemma 2.1.4).

Put $\mathcal{D} := \{(x,y) \mapsto \sum_{ij} U_x A_i(x) B_j(y)^* U_y^* \mid A_1, \ldots, A_k, B_1, \ldots, B_l :$ $(0,\infty) \to E_\alpha(t_0)$ strongly continuous of compact support $\}$ [5]. Then $\mathcal{D}$ is dense. Let $F(x,y) = \sum_{ij} U_x A_i(x) B_j(y)^* U_y^*$ and consider $\alpha_\lambda(U(x-\lambda, s-\lambda))$ $(0 < \lambda < x, s)$. Because α is unital $\lambda \mapsto \alpha_\lambda(U(x-\lambda, s-\lambda))$ is a weakly hence strongly continuous family of unitaries in all parameters on H s.t. $\alpha_\lambda(U(x-\lambda, s-\lambda))E_\alpha(s) = E_\alpha(x)$ and acts like $\mathbf{1}_\lambda \otimes U(x-\lambda, s-\lambda)$ ($T \in E_\alpha(\lambda)$ and $S \in E_\alpha(s-\lambda)$ implies $\alpha_\lambda(U(x-\lambda, s-\lambda))TS = TU(x-\lambda, s-\lambda)S$). Thus we have "extended" $\mathbf{1}_\lambda \otimes U(x-\lambda, s-\lambda)$ to a unitary in $\mathcal{B}(H)$. Now

$$\alpha_\lambda(U(x-\lambda, s-\lambda))U_s \xrightarrow{s \to x} U_x$$

strongly. Therefore $(\mathbf{1}_\lambda \otimes U(x-\lambda, s-\lambda))F(s,y) \to F(x,y)$ for $s \searrow x$ in norm because $\sum_{i,j} \alpha_\lambda(U(x-\lambda, s-\lambda))U_s A_i(s) B_j(y)^* U_y^* \to \sum_{i,j} U_x A_i(x) B_j(y)^* U_y^*$ strongly.

(iii): It follows from (ii) that $\tilde{F}^\varepsilon$ defined by

$$\tilde{F}^\varepsilon(x,y) = \tfrac{1}{\varepsilon^2} \int_x^{x+\varepsilon} ds \int_{max(0,x-\varepsilon)}^x d\lambda \, (U(x-\lambda, s-\lambda) \otimes \mathbf{1}_\lambda) F(s,y)$$

converges pointwise to F and has an L^1-majorant (because $(s,y) \mapsto \|F(s,y)\| \in C_c(\mathbb{R}_+ \times \mathbb{R}_+)$ we may take $(x,y) \mapsto \sup_{s \in [x, x+\varepsilon)} \|F(s,y)\|$). This is also a majorant for $F^{N,\varepsilon}$ and $(P_N(x-\lambda, s-\lambda) \otimes \mathbf{1}_\lambda)F(s,y) \to$

---

[4] Unfortunately this does not match with the identification $U_x : E(x) \to E(t_0)$ above

[5] equivalently norm continuous into the Hilbert space $E_\alpha(t_0)$

$(U(x-\lambda, s-\lambda) \otimes 1_\lambda) F(x,y)$ in norm implies $F^{N,\varepsilon} \xrightarrow{N\to\infty} \tilde{F}^\varepsilon$ in L^1 (Appendix Thm.2.(ii)). Now $\tilde{F}^\varepsilon \xrightarrow{\varepsilon\to 0} F$ and we can find the required sequences ε_l, N_l. □

Let $N_\varepsilon := \{(\mu, \mu+\lambda) | \mu, \lambda \in (0,\varepsilon)\} \subseteq \mathbb{R}_+ \times \mathbb{R}_+$.

Proposition 4.2.14 $L^1(K_E)$ has an approximate unit [6] given by

$$K_{N_l}^{\varepsilon_l}(x,y) = \begin{cases} \frac{1}{\varepsilon_l^2} \sum_{n=0}^{N_l} e_n(x) \otimes \overline{e_n(y)} & \text{if } (x,y) \in N_{\varepsilon_l} \\ 0 & \text{otherwise} \end{cases}$$

for a certain sequence of natural numbers (N_l) and a certain zero sequence (ε_l).

<u>Proof:</u> $\|K_N^\varepsilon\|_1 = 1$ for any $\varepsilon > 0$. Let $F \in \mathcal{D} \subseteq L^1(K_E)$ and consider for $\varepsilon > 0$

$$\begin{aligned}(K_N^\varepsilon F)(x,y) &= \int_0^\infty ds \Big[\int_0^{min(s,y)} d\lambda \; K_N^\varepsilon(x,s)(F(s-\lambda, y-\lambda) \otimes 1_\lambda) \\ &+ \int_0^{min(x,s)} d\lambda \; (K_N^\varepsilon(x-\lambda, s-\lambda) \otimes 1_\lambda) F(s,y) \Big]\end{aligned}$$

Let $c > 0$ be as in 4.2.13.(iii) and $2\varepsilon < c$. In this case the first term in the product vanishes and we are left with

$$\int_0^\infty ds \int_0^{min(x,s)} d\lambda \; (K_N^\varepsilon(x-\lambda, s-\lambda) \otimes 1_\lambda) F(s,y)$$

But by the definition of N_ε, this is the expression in 4.2.13.(iii). Let $\{F_k | k \in \mathbb{N}\} \subseteq \mathcal{D}$ be a countable dense sequence in the unitsphere of $L^1(K_E)$. There are sequences $\varepsilon_l \to 0$ and $N_l \to \infty$ s.t. $\|K_{N_l}^{\varepsilon_l} F_1 - F_1\|, \ldots, \|K_{N_l}^{\varepsilon_l} F_l - F_l\| < \frac{1}{l}$. Thus $K_{N_l}^{\varepsilon_l}$ is an approximate unit. □

The set $L^1(E)$ of L^1-sections of E with the product

$$(f * g)(x) := \int_0^x dt \; f(t) g(x-t)$$

is a nonselfadjoint Banach algebra.

---

[6] By approximate unit we always mean two sided and bounded by one. If E contains a unit (the only case we need later on), then there are easier approximate units

ARVESON'S SPECTRAL C*-ALGEBRAS

Proposition 4.2.15 *Let E be a product system containing at least one unit u* [7]. *Define a semigroup T on the Banach space $L^1(E)$ as follows*

$$(T(t)\xi)(s) = \begin{cases} u(t)\xi(s-t) & \text{if } s > t \\ 0 & \text{otherwise} \end{cases}$$

Then T is a strongly continuous semigroup of isometries.

Proof: $t \mapsto T(t)\xi$ is measurable for each $\xi \in L^1(E)$ because $t \mapsto u(t) \in E$ is measurable and the multiplication in E is measurable. Furthermore

$$\|T(t)\xi\| = \int_t^\infty dx \underbrace{\|u(t)\xi(x-t)\|}_{=\|u(t)\|\|\xi(x-t)\|} = \int_0^\infty dx \,\|\xi(x)\|$$

shows that $T(t)$ is isometric.

We proceed as in [Ar 89a]. By [HiPh 10.2.3], T is strongly continuous and by [HiPh 10.5.1], $T(t)\xi \to \xi$ for $t \to 0$ follows if we can show

$$\overline{\bigcup_{t>0} T(t)(L^1(E))} = L^1(E)$$

Now let (e_n) be a trivializing sequence for E, $(f_l) \subseteq L^1(\mathbb{R}_+)$ a dense sequence and (t_k) a sequence of positive numbers converging to zero. Let fe be the section $t \mapsto f(t)e(t)$. The set $\{T(t_k)f_l e_n | k, l, n \in \mathbb{N}\}$ is generating in the sense of Appendix Prop.7. Therefore it spans a dense subspace of sections in $L^1(E)$. $\square$

Corollary 4.2.16 *For any family s.t. $f_\varepsilon \in L^1(\mathbb{R}_+)$ is positive, $\mathrm{supp}(f_\varepsilon) \subseteq (0, \varepsilon)$ and $\int_0^\varepsilon dt\, f_\varepsilon(t) = 1$, the sections $f_\varepsilon u$ given by $t \mapsto f_\varepsilon(t)u(t)$ form a two sided approximate unit in $L^1(E)$ for $\varepsilon \to 0$.*

Proof: Let f_ε be as in the assumption. Observe that for each strongly continuous semigroup of isometries $S(t)$ on a Banach space X s.t. $S(t)x \xrightarrow{t \to 0} x$ we have $\int_0^\varepsilon dt\, f_\varepsilon(t) S(t)x \xrightarrow{\varepsilon \to 0} x$. Thus $(f_\varepsilon u) * g = \int_0^\varepsilon dt\, f_\varepsilon(t) T(t) g \xrightarrow{\varepsilon \to 0} g$ for any $g \in L^1(E)$. Defining a similar semigroup by multiplying with $u(t)$ from the right, $f_\varepsilon u$ is also a right approximate unit. $\square$

---

[7] normalized i.e. $\|u(t)\| = 1\ \forall t > 0$.

4.2.3 $C^*(E)$ and its Universal Property

For any representation $\phi : E \to \mathcal{B}(H)$ and $f \in L^1(E)$ we may define the operator $\phi(f)\xi := \int_0^\infty dt\, \phi(f(t))\xi$ on H. $\phi(f(t))^*\phi(f(t)) = \|f(t)\|^2 \mathbf{1}$ implies $\|f(t)\| = \|\phi(f(t))\|$ and therefore $\|\phi(f)\| \leq \int_0^\infty dt\, \|\phi(f(t))\| = \int_0^\infty dt\, \|f(t)\|$. In particular, $\phi(f)$ is bounded. For $v \in E(t)$ we have the section

$$(vf)(x) = \begin{cases} vf(x-t) & \text{if } x > t \\ 0 & \text{otherwise} \end{cases}$$

Some of the main results in [Ar 90a] may be summarized as follows. (They will be used together with Prop.4.2.8 and 4.2.15 to show Prop.4.2.19 and 4.2.20.)

Theorem 4.2.17 *(i) For each representation $\pi : C^*(E) \to \mathcal{B}(H)$, there is a unique representation $\phi_\pi : E \to \mathcal{B}(H)$ s.t. for $f, g \in L^1(E)$ and $v \in E$:*

$$\phi_\pi(v)\pi(f \otimes \bar{g}) = \pi(vf \otimes \bar{g})$$

(ii) For each representation $\phi : E \to \mathcal{B}(H)$, there is a unique nondegenerate representation $\pi_\phi : C^*(E) \to \mathcal{B}(H)$ s.t. for $f, g \in L^1(E)$:

$$\pi_\phi(f \otimes \bar{g}) = \phi(f)\phi(g)^*$$

(iii) $\pi_{\phi_\pi} = \pi$ for any nondegenerate representation π of $C^*(E)$ and $\phi_{\pi_\phi} = \phi$ for any representation ϕ of E.

Proof: (i) [Ar 90a Prop.3.4].

(ii) [Ar 90a Thm.2.16].

(iii) $\pi_{\phi_\pi} = \pi$ is [Ar 90a Cor.1 of Thm.3.4] and $\phi_{\pi_\phi} = \phi$ follows from $\phi(v)\phi(f)\phi(g)^* = \phi(vf)\phi(g)^*$, $v \in E$, $f, g \in L^1(E)$ for any representation $\phi : E \to \mathcal{B}(H)$ and uniqueness in (i). □

Remark 4.2.18 *(i) If $E = (0, \infty) \times \mathbb{C}$ is the trivial product system (c.f. the remark after Def.4.1.3), let ϕ be a representation and s be the section identically equal to 1. Put $V(t) := \phi(s(t))$. Then $V(t)$ is measurable and $V(x + y) = V(x)V(y)$. Thus $V(t)$ is a strongly continuous*

ARVESON'S SPECTRAL C*-ALGEBRAS

semigroup of isometries. It follows that $C^*(E) = \mathcal{W}_0$ is the Wiener-Hopf algebra.

(ii) Suppose $E = E_\alpha$ for some e_0-semigroup α (which is no restriction). Then for each representation $\pi : C^*(E) \to \mathcal{B}(H)$ there exists an e_0-semigroup β on H s.t. $E_\beta \cong E_\alpha$.

(iii) A representation π of $C^*(E)$ is called singular (essential) if the corresponding representation of E is singular (essential). The representation $\lambda : C^*(E) \to \mathcal{B}(L^2(E))$ associated to the regular representation $l : E \to \mathcal{B}(L^2(E))$ (c.f.Ex.4.1.4) is called the Fock representation. Let γ_t be the strongly continuous group on $C^*(E)$ defined by $\gamma_t(K)(x,y) := e^{it(x-y)}K(x,y)$ for $K \in L^1(K_E)$. γ is called the gauge group and is unitarily implemented in λ via $(U_t\xi)(x) = e^{itx}\xi(x)$ s.t. the generator has positive spectrum. $C_r^*(E) := \lambda(C^*(E))$ is called the reduced spectral algebra, and it follows [PeTa 79] that the crossed product $\mathbb{R} \rtimes C_r^*(E)$ is necessarily nonsimple. In [Ar 90a sec.7] $\|\lambda(x)\| \leq \sup_{t \in \mathbb{R}} \|\pi \circ \gamma_t(x)\|$ is shown for each representation π of $C^*(E)$ and this is the key to Arveson's proof of the simplicity of $C_r^*(E)$ for product systems admitting at least one unit.

(iv) Thm.4.2.17 can also be used to show that $C^*(E)$ is always nuclear [Ar 90a sec.4].

Proposition 4.2.19 Let E be a product system containing at least one unit and $\pi : C^*(E) \to \mathcal{B}(H)$ be a nondegenerate representation of E. Then $\phi_\pi(L^1(E)) \subseteq \pi(C^*(E))$ where ϕ_π is as in Thm.4.2.17.(i).

Proof: Let ϕ_π be as in Thm.4.2.17.(i) and u be a normalized unit in E. Then $V(t) := \phi_\pi(u(t))$ is a strongly continuous semigroup of isometries on H s.t. for $F \in L^1(\mathbb{R}_+^2)$ the integrals $\int_0^\infty dx \int_0^\infty dy\, F(x,y)V(x)V(y)^*$ lie in $\pi(C^*(E))$, and Prop.4.2.8 implies $\int_0^\infty dx\, f(x)V(x) \in \pi(C^*(E))$ for each $f \in L^1(\mathbb{R}_+)$. Denote as before the section $t \mapsto f(t)u(t)$ by fu. Consider for $\xi \in L^1(E)$ and $\varepsilon > 0$

$$\pi(\xi \otimes \overline{\tfrac{1}{\varepsilon}\chi_{[0,\varepsilon)}u}) = \phi_\pi(\xi)\phi_\pi(\tfrac{1}{\varepsilon}\chi_{[0,\varepsilon)}u)^* \in \pi(C^*(E))$$

according to Thm.4.2.17.(ii). Therefore

$$\pi(\xi \otimes \overline{\tfrac{1}{\varepsilon}\chi_{[0,\varepsilon)}u})\phi_\pi(\tfrac{1}{\varepsilon}\chi_{[\varepsilon,2\varepsilon)}u) = \tfrac{1}{\varepsilon^2}\phi_\pi(\xi)\phi_\pi(\chi_{[0,\varepsilon)}u)^*\phi_\pi(\chi_{[\varepsilon,2\varepsilon)}u)$$
$$= \tfrac{1}{\varepsilon^2}\phi_\pi(\xi)\int_0^\varepsilon dt \int_\varepsilon^{2\varepsilon} ds\, V(s-t)$$

Substituting $s - t = r$ the last integral becomes (including $\tfrac{1}{\varepsilon^2}$)

$$\int_0^{2\varepsilon} dr\, f_\varepsilon(r) V(r)$$

with $f_\varepsilon : \mathbb{R}_+ \to \mathbb{R}_+$ the function

$$f_\varepsilon(r) = \begin{cases} \tfrac{1}{\varepsilon^2} r & \text{if } r \in (0,\varepsilon) \\ \tfrac{1}{\varepsilon^2}(2\varepsilon - r) & \text{if } r \in [\varepsilon, 2\varepsilon) \\ 0 & \text{otherwise} \end{cases}$$

It follows $\int_0^\infty dr\, f_\varepsilon(r) = \tfrac{2}{\varepsilon^2}\int_0^\varepsilon dr\, r = 1$. Hence the above expression is equal to

$$\phi_\pi(\xi)\phi_\pi(f_\varepsilon u) = \phi_\pi(\xi * f_\varepsilon u) \xrightarrow{\varepsilon \to 0} \phi_\pi(\xi)$$

in norm using Cor.4.2.16. $\square$

Proposition 4.2.20 *For any product system E having at least one unit, E is contained in the multiplier algebra of $C^*(E)$.*

<u>Proof:</u> Let $v \in E(t) \subseteq E$. We show that for $f, g \in L^1(E)$ there are sections $f_1, g_1 \in L^1(E)$ s.t.

$$\pi(f \otimes \bar{g})\phi_\pi(v) = \phi_\pi(f)\phi_\pi(g_1)^* + \phi_\pi(f_1)$$

This implies

$$\pi(\mathcal{F})\phi_\pi(v) \subseteq \pi(\mathcal{F}) + \phi_\pi(L^1(E))$$

where $\mathcal{F}$ denotes the finite rank elements in $L^1(K_E)$. Because $\pi(C^*(E)) = \overline{\pi(\mathcal{F})}$ and $\phi_\pi(v)$ is bounded

$$\pi(C^*(E))\phi_\pi(v) = \overline{\pi(\mathcal{F})}\phi_\pi(v) \subseteq \overline{\pi(\mathcal{F})\phi_\pi(v)}$$

$$\subseteq \overline{\pi(\mathcal{F}) + \phi_\pi(L^1(E))} = \pi(C^*(E))$$

where the last equality holds by Prop.4.2.19.

Because $\phi_\pi(v)\pi(\mathcal{F}) \subseteq \pi(\mathcal{F})$ (Thm.4.2.17.(i)) $\phi_\pi(v)$ is a two sided multiplier.

To find f_1 and g_1, write for $0 < s < t$ and (e_n) a trivializing sequence of sections in E:

$$v = \sum_{n,m} \alpha_{n,m}(s) e_n(s) e_m(t-s)$$

with $s \mapsto \alpha_{n,m}(s)$ measurable on $(0,t)$ and $\sum_{n,m} |\alpha_{n,m}(s)|^2 = \|v\|^2$. Define f_1 to be the section

$$f_1(y) = \sum_{n,m} \int_{max(t-y,0)}^{t} ds\, \alpha_{n,m}(s) \langle g(s), e_n(s) \rangle f(y-t+s) e_m(t-s)$$

f_1 is a measurable section and

$$\|f_1(y)\| \leq \int_{max(t-y,0)}^{t} ds\, \underbrace{\|\sum_{n,m} \alpha_{n,m}(s) \langle g(s), e_n(s) \rangle e_m(t-s)\|}_{=\|g(s)\|^{-1}\|(\Theta_{g(s),g(s)} \otimes 1)v\|} \|f(y-t+s)\|$$

$$\leq \int_{max(t-y,0)}^{t} ds\, \|v\| \|g(s)\| \|f(y-t+s)\|$$

Moreover, the last expression is a majorant for the finite subsums in n and m. If we put

$$f_t(z) = \begin{cases} \|f(z-t)\| & \text{if } z > t \\ 0 & \text{otherwise} \end{cases}$$

then $f_t \in L^1(\mathbb{R}_+)$ and we obtain

$$\int_0^\infty dy \, \|f_1(y)\| \leq \int_0^\infty dy \int_{max(t-y,0)}^t ds \, \|v\| \|g(s)\| \|f(y-t+s)\|$$

$$\leq \int_0^\infty dy \int_0^t ds \, \|v\| \|g(s)\| f_t(y+s) < \infty$$

This shows that $f_1 \in L^1(E)$. For $y > t$ we write

$$g(y) = \sum_{n,m} \beta_{n,m}(y) e_n(t) e_m(y-t)$$

with $\beta_{n,m} : (t, \infty) \to \mathbb{C}$ measurable and $\sum_{n,m} |\beta_{n,m}(y)|^2 = \|g(y)\|^2$. Define for $z \in \mathbb{R}_+$

$$g_1(z) := \sum_{n,m} \beta_{n,m}(t+z) \langle v, e_n(t) \rangle e_m(z)$$

Then g_1 is measurable and by the Cauchy-Schwarz inequality

$$\|g_1(z)\|^2 = \sum_m \left| \sum_n \beta_{n,m}(t+z) \langle v, e_n(t) \rangle \right|^2 \leq \|g(t+z)\|^2 \|v\|^2$$

which shows that $g_1 \in L^1(E)$. Now

$$\pi(f \otimes \bar{g}) \phi_\pi(v) = \phi_\pi(f) \phi_\pi(g)^* \phi_\pi(v)$$

$$= \phi_\pi(f) \int_0^t ds \, \phi_\pi(g(s))^* \phi_\pi(v) + \phi_\pi(f) \int_t^\infty ds \, \phi_\pi(g(s))^* \phi_\pi(v)$$

Because $\|f_1\|$ and $\|g_1\|$ are also majorants for finite subsums over n and m we may interchange sums and integrals. Thus

$$\pi(f \otimes \bar{g}) \phi_\pi(v) = \sum_{n,m} \phi_\pi(f) \int_0^t ds \, \alpha_{nm}(s) \langle g(s), e_n(s) \rangle \phi_\pi(e_m(t-s)) +$$

$$+ \sum_{n,m} \phi_\pi(f) \int_t^\infty ds \, \overline{\beta_{nm}(s)} \langle e_n(t), v \rangle \phi_\pi(e_m(s-t))^*$$

$$= \phi_\pi(f_1) + \phi_\pi(f) \phi_\pi(g_1)^*$$

which means that f_1 and g_1 are as required. $\square$

Remark 4.2.21 *Let E be a product system with a unit and π a faithful nondegenerate essential representation of $C^*(E)$ on H with corresponding E_0-semigroup α. Any automorphism φ of $C^*(E)$ extends to the multiplier algebra $\mathcal{M}(C^*(E))$ and we can define $U_t^\varphi = \sum_{n=0}^\infty \varphi(e_n(t))e_n(t)^*$ as an operator in $\mathcal{B}(H)$. (U_t^φ can be defined for any product system as an element in $C^*(E)^{**}$. This follows from [Ar 90a Thm.3.4]). Then $U_t^\varphi \alpha_t(U_s^\varphi) = U_{t+s}^\varphi$ and we obtain a (possibly nonunitary) cocycle for α. Thus there is a close connection between cocycle conjugacy and automorphisms of $C^*(E)$. For instance $U_t^\varphi := e^{ist}$ with $s \in \mathbb{R}$ defines the gauge automorphism γ_s. Conversely, if we have a strictly continuous cocycle of the E_0-semigroup α consisting of unitary multipliers, then $\varphi_U(\phi_\pi(f)) = \int_0^\infty dt\, U_t \phi_\pi(f(t))$ defines an endomorphism of $C^*(E)$. This is in analogy to Cuntz's analysis of automorphisms of $\mathcal{O}_n$ [Cu 78] where the cocycle corresponds to the sequence $U, U\Phi(U), U\Phi(U)\Phi^2(U), \ldots$, Φ the canonical endomorphism of $\mathcal{O}_n$ and U a unitary in $\mathcal{O}_n$.*

Remark 4.2.22 *One could try to define a spectral algebra of a tensor decomposition as being generated by the intertwining isometries (or integrals over families of such isometries) of the associated family of endomorphisms. However, there seems to be some ambiguity in the definition because of the ambiguity of the family.*

4.2.4 The C^*-Algebras $\mathcal{W}_n$

The following is a minor improvement of the claim and the proof in [Ar 91] (which he just overlooked). It is shown that the spectral algebra $C^*(E_n)$ can also be generated by integrals over semigroups of isometries $U_i(t)$. Thereby we also obtain projections in $\mathcal{W}_n$. Let H be a separable Hilbert space and $(U_i(t))_{i=0,1,\ldots}$ a (possibly infinite) sequence of semigroups of isometries s.t. $U_i(t)^* U_j(t) = e^{-\lambda_{ij} t}$ with (λ_{ij}) a matrix of conditional rank n for one (and hence each (c.f. Appendix Prop.15)) $t > 0$, i.e. $\{U_i(t) | i = 0, \ldots\}$ generates a Hilbert space of dimension $n+1$. Consider $\mathcal{W}_n := C^*(\mathcal{S}) := C^*(\{U_i(f) | i = 0, \ldots f \in L^1(\mathbb{R}_+)\})$ where $U_i(f) = \int_0^\infty dt\, f(t) U_i(t)$. Then it turns out that $\mathcal{W}_n \cong C^*(E_n)$.

In $C^*(\mathcal{S})$ there is the dense $*$-subalgebra

$$\mathcal{P} := \Big\{ \int dt_1 \ldots dt_p ds_1 \ldots ds_q \, f(t_1, \ldots, t_p; s_1, \ldots, s_q) U_{i_1}(t_1) \ldots$$
$$\ldots U_{i_p}(t_p) U_{j_q}(s_q)^* \ldots U_{j_1}(s_1)^* | \, i_r, j_s = 0, \ldots \, f \in L^1(\mathbb{R}_+^{p+q}) \Big\}$$

and the subalgebra

$$\mathcal{P}_+ := \Big\{ \int dt_1 \ldots dt_p f(t_1, \ldots, t_p) U_{i_1}(t_1) \ldots U_{i_p}(t_p) | i_r = 0, \ldots f \in L^1(\mathbb{R}_+^p) \Big\}$$

If $S_p = \{(s_1, \ldots, s_p) | 0 \leq s_1 \leq \ldots \leq s_p\} \subseteq \mathbb{R}_+^p$ with the Lebesgue measure as before, then each element in $\mathcal{P}_+$ can be written as

$$\int_0^\infty ds_p \int_0^{s_p} ds_{p-1} \ldots \int_0^{s_2} ds_1 \, f(s_1, \ldots, s_p) U_{i_1}(s_1) U_{i_2}(s_2 - s_1) \ldots U_{i_p}(s_p - s_{p-1})$$

where f is in $L^1(S_p)$.

Now let $H(\Lambda)$ be the conditional Hilbert space to the matrix $\Lambda = (\lambda_{ij})$ which is the completion of $\{f : \{0, 1, \ldots\} \to \mathbb{C} \mid suppf$ finite and $\sum_i f(i) = 0\}$ with scalar product $\sum \lambda_{ij} \overline{f(i)} g(i)$. Let $f = \delta_i$, $\delta_i(j) := \delta_{ij}$. Then the set $\{\delta_0 - \delta_i | i = 1, \ldots\}$ is total in $H(\Lambda)$. We denote the element $\delta_0 - \delta_i$ (or more precisely its image) in $H(\Lambda)$ by ξ_i.

Lemma 4.2.23 *The set* $S = \{\sum_{k=1}^l \chi_{[a_k,b_k)} \otimes \xi_{i_k} | \, 0 < a_1 < b_1 < \ldots < b_l$ *and* $i_k \in \{1, \ldots\}\}$ *is strongly spanning in* $L^2(\mathbb{R}_+, H(\Lambda))$

Proof: Immediate by Lemma 2.1.19. □

Lemma 4.2.24 *For each* $s \in S$ *let* $s_t = s|(0, t)$ *and* $f \in L^1(\mathbb{R}_+)$. *Consider sections of the form* $t \mapsto f(t) exp(s_t) \in E_n(t)$. *Then the span of all of them is dense in* $L^1(E_n)$.

Proof: Let $S_0 := \{\sum_{k=1}^l \chi_{[a_k,b_k)} \otimes \xi_{i_k} \in S | \, a_k, b_k \in \mathbb{Q}_+\}$. Then S_0 is countable and still strongly spanning. The set $\{t \mapsto exp(s_t) | s \in S_0\}$ is generating in the sense of Appendix Prop.7 which implies the claim. □

Theorem 4.2.25 *There exists a representation* ϕ *of* E_n *and by integration of* $L^1(E_n)$ *on* H *where* $n = \dim H(\Lambda)$ *s.t.*

$$\phi(e^{-\frac{tx_i}{2}} exp(\chi_{[0,t)} \otimes \xi_i)) = U_i(t)$$

ARVESON'S SPECTRAL C^*-ALGEBRAS

$i = 1, \ldots,\ x_i = (\lambda_{00} + \lambda_{ii} - \lambda_{0i} - \lambda_{i0})^2 = \|\xi_i\|^2,$

$$\phi(exp(0|[0,t))) = U_0(t)$$

and $\phi(L^1(E_n)) \subseteq C^*(\mathcal{S})$.

<u>Proof:</u> If $s \in S$, $s = \sum_{i=1}^{l} \chi_{[a_{2i-1}, a_{2i})} \otimes \xi_{k_{2i}}$, $0 < a_1 < \ldots < a_{2l}$, $s_t = s|(0,t)$, $c_t = e^{\frac{\|s_t\|^2}{2}}$, define for $exp(s_t) \in E_n(t)$

$$\phi(exp(s_t)) := c_t\, U_{k_1}(a_1) U_{k_2}(a_2 - a_1) \ldots U_{k_r}(t - a_{r-1})$$

where $r \in \mathbb{N}$ is the unique number s.t. $t \in (a_{r-1}, a_r]$, k_{2i} comes from s and $k_{2i+1} = 0$ for $i = 0, \ldots, l$. One checks that

$$\phi(exp(s_t))^* \phi(exp(s_t')) = e^{\langle s_t, s_t' \rangle} \mathbf{1}$$

for $s, s' \in S$ and

$$\phi(exp(s_t)) \phi(exp(s_r')) = \phi(exp(s_t + S_t s_r'))$$

Finally $t \mapsto \phi(exp(s_t))$ is measurable. We obtain a representation of E_n and thus by integration also one of $L^1(E_n)$.

We claim that for each $\alpha \in L^1(\mathbb{R}_+)$

$$\int_0^\infty dt\, \alpha(t) \phi(exp(s_t)) \in C^*(\mathcal{S})$$

Put for $p \leq 2l$ and ε sufficiently small

$$f_p^\varepsilon(s_1, \ldots, s_p) = \begin{cases} \frac{1}{\varepsilon^p} & \text{if } a_i - s_i \in (0, \varepsilon) \\ 0 & \text{otherwise} \end{cases}$$

By strong continuity of the U_i, we can see that

$$\sum_{p=1}^{2l} \int_0^\infty dt \int_0^t ds_p \ldots \int_0^{s_2} ds_1\, \alpha(t)\, c_t\, f_p^\varepsilon(s_1, \ldots, s_p)\, U_{k_1}(s_1) U_{k_2}(s_2 - s_1)$$

$$\ldots U_{k_p}(s_p - s_{p-1}) U_{k_{p+1}}(t - s_p) \xi \xrightarrow{\varepsilon \to 0} \int_0^\infty dt\, \alpha(t) \phi(exp(s_t)) \xi$$

for each $\xi \in H$ because the integral converges pointwise and has an integrable majorant. For $t > 0$ put

$$h_t := \left[\{ U_{i_1}(t_1) \ldots U_{i_k}(t_k) | t_1 + \ldots + t_k = t\ ,\ i_l \in \{0, \ldots\} \} \right]$$

We obtain a Hilbert space in $\mathcal{B}(H)$. Hence the strong and the norm topology coincide on h_t and the integral converges even in norm. Thus $\phi(L^1(E_n)) \subseteq C^*(\mathcal{S})$ by 4.2.24. □

We proved the following:

Theorem 4.2.26 $C^*(\mathcal{S}) \cong C^*(E_n)$ where $n = \dim H(\Lambda)$.

Using Prop.4.2.4, we compute isometric generators of $\mathcal{W}_n^\sim$:

Lemma 4.2.27 *Suppose $\mathcal{W}_n$ is generated by m semigroups of isometries U_i with the relation $U_i(t)^* U_j(t) = e^{-\lambda_{ij} t} \mathbf{1}$. Then $\mathcal{W}_n^\sim$ is generated by isometries s_i, $i = 1, \ldots, m$ subject to the relations*

$$s_i^* s_j = \frac{2-\lambda_{ij}}{\lambda_{ij}+2}\mathbf{1} + \frac{\lambda_{ij}}{\lambda_{ij}+2}(s_i^* + s_j)$$

Proof: We put $s_i := \mathbf{1} - 2\int_0^\infty dt\, e^{-t} U_i(t)$ which is the isometry considered in sec.4.2.1. Because $C^*(s_i) = C^*(\{\int_0^\infty dt\, f(t) U_i(t) | f \in L^1(\mathbb{R}_+)\})^\sim$ we have $C^*(s_i | i = 0, \ldots, m) = \mathcal{W}_n^\sim$. For the relation we have to compute:

$$s_i^* s_j = \mathbf{1} - 2\int_0^\infty dt\, e^{-t} U_i(t)^* - 2\int_0^\infty dt\, e^{-t} U_j(t)$$

$$+ 4\int_0^\infty dt \int_0^\infty ds\, e^{-t-s} U_i(t)^* U_j(s)$$

The last term is

$$\int_0^\infty dt \int_0^\infty ds\, e^{-t-s} U_i(t)^* U_j(s) =$$

$$= \int_0^\infty dt \int_0^t ds\, e^{-t-s-\lambda_{ij} s}[U_i(t-s)^* + U_j(t-s)]$$

substituting $t - s = r$ and using s and r as new coordinates, this is equal to:

$$= \int_0^\infty ds\, e^{-(\lambda_{ij}+2)s} \int_0^\infty dr\, e^{-r}(U_i(r)^* + U_j(r))$$

A little computation gives the desired constants. □

Corollary 4.2.28 $\mathcal{W}_n$ *contains nontrivial projections.*

Proof:
$$2\left(\int_0^\infty dt\, e^{-t}U_i(t) + \int_0^\infty dt\, e^{-t}U_i(t)^* - 2\int_0^\infty dt \int_0^\infty ds\, e^{-t-s}U_i(t)U_i(s)^*\right)$$
is a nontrivial projection in $\mathcal{W}_n$. $\square$

4.3 $C^*(E_n)$ as a Crossed Product

4.3.1 The Banach Algebra Crossed Product $L^1(\mathbb{R}, L^1(T_n))$

We can describe the action α on $\mathcal{A}_n$ considered in Prop.3.3.22.(i) directly on $L^1(T_n) = L^1(T_n(\mathbb{R}))$ and form the Banach algebra crossed product. By the existence of approximate units in $L^1(T_n)$, it follows that the enveloping C^*-algebra of the crossed product is equal to $\mathbb{R} \rtimes \mathcal{A}_n$.

In $\mathcal{H} = \mathcal{F}^s(L^2(\mathbb{R}, \mathbb{C}^n))$ consider the unitary $U(r) = exp(S_r)$, $r \in \mathbb{R}$, S_r the twosided shift. S_r is also a unitary from $L^2((-\infty, t), \mathbb{C}^n)$ to $L^2((-\infty, t+r), \mathbb{C}^n)$. Therefore we may view $U(r)$ as a unitary from $\mathcal{H}_t$ to $\mathcal{H}_{t+r} = \mathcal{F}^s(L^2((-\infty, t+r), \mathbb{C}^n))$ and denote it also by $U(r)$. We illustrate this slight abuse of notation:

Remark 4.3.1 $U(r)A_x U(r)^* \in \mathcal{B}(\mathcal{H}_{x+r})$ *for* $A_x \in \mathcal{B}(\mathcal{H}_x)$ *viewing* $U(r)$ *as a map from* $\mathcal{H}_x$ *to* $\mathcal{H}_{x+r}$. *If* $A_x = A'_{x'} \otimes \mathbf{1}_{[x',x)}$, *then* $U(r)(A'_{x'} \otimes \mathbf{1}_{[x',x)})U(r)^* = U(r)A'_{x'}U(r)^* \otimes \mathbf{1}_{[x'+r,x+r)}$ *where on the l.h.s. we view* $U(r)$ *as a unitary from* $\mathcal{H}_x$ *to* $\mathcal{H}_{x+r}$ *and on the r.h.s. as a unitary from* $\mathcal{H}_{x'}$ *to* $\mathcal{H}_{x'+r}$.

Proof: It may be checked on exp-vectors. $\square$

Now let $(K_t) \in L^1(T_n(\mathbb{R})) = L^1(T_n)$ be any L^1-section and define for $r \in \mathbb{R}$ the section $(K_t^{\alpha_r})$ by
$$K_t^{\alpha_r} := U(r)K_{t-r}U(r)^* \in \mathcal{K}(\mathcal{H}_t)$$
using the above identification.

Proposition 4.3.2 α_r *is an isometric *-automorphism of* $L^1(T_n)$.

<u>Proof:</u> It is immediate to check $\|\alpha_r(K)\|_1 = \|K\|_1$. To show multiplicativity we compute $\alpha_r(KL)$ for any $K, L \in L^1(T_n)$. The value of this section at $t \in \mathbb{R}$ is

$$U(r) \left[\int_{-\infty}^{t-r} ds \left[K_{t-r}(L_s \otimes \mathbf{1}) + (K_s \otimes \mathbf{1})L_{t-r} \right] \right] U(r)^* =$$

$$= U(r) \left[\int_{-\infty}^{t} ds \left[K_{t-r}(L_{s-r} \otimes \mathbf{1}) + (K_{s-r} \otimes \mathbf{1})L_{t-r} \right] \right] U(r)^*$$

$$= \int_{-\infty}^{t} ds \left[U(r)(K_{t-r}(L_{s-r} \otimes \mathbf{1}))U(r)^* + U(r)((K_{s-r} \otimes \mathbf{1})L_{t-r})U(r)^* \right]$$

$$= \int_{-\infty}^{t} ds \left[U(r)K_{t-r}U(r)^*U(r)(L_{s-r} \otimes \mathbf{1})U(r)^* + \right.$$
$$\left. + U(r)(K_{s-r} \otimes \mathbf{1})U(r)^*U(r)L_{t-r}U(r)^* \right]$$

$$= \int_{-\infty}^{t} ds \left[K_t^{\alpha_r}(L_s^{\alpha_r} \otimes \mathbf{1}) + (K_s^{\alpha_r} \otimes \mathbf{1})L_t^{\alpha_r} \right]$$

where we have used the above mentioned abuse of notation several times. □

Now consider a Banach-*-algebra dynamical system $(\mathbb{R}, \alpha, B)$ given by a Banach-*-algebra B with a strongly continuous action α (i.e. $\lim_{t\to 0} \|\alpha_t(x) - x\| = 0 \quad \forall x \in B$) of $\mathbb{R}$ by isometric *-automorphisms and equip $L^1(\mathbb{R}, B)$ with the product:

$$(x * y)(t) := \int_{\mathbb{R}} ds \, x(s)\alpha_s(y(t-s))$$

$x, y \in L^1(\mathbb{R}, L^1(T_n))$ and involution

$$x^*(t) := \alpha_t(x(-t)^*)$$

We need a few straightforward generalizations of well known assertions for C^*-algebras to Banach-*-algebras with bounded approximate units.

ARVESON'S SPECTRAL C^*-ALGEBRAS

Lemma 4.3.3 *Let B be a Banach-$*$-algebra with bounded approximate unit (e_i) and (R, L) a double centralizer i.e. a pair of maps $R, L : B \to B$ s.t. $R(xy) = xR(y)$, $L(xy) = L(x)y$ and $xL(y) = R(x)y$ for any $x, y \in B$. Then R and L are norm bounded linear maps. Moreover, for each nondegenerate representation $\pi : B \to \mathcal{B}(H)$ there is a unique extension of π to $\bar{\pi} : DC(B) \to \mathcal{B}(H)$ with $DC(B)$ the algebra of double centralizers.*

Proof: By [HeRo II (32.49)] L and R are bounded linear maps. The proof for the existence and uniqueness of $\bar{\pi}$ is as for a C^*-algebra : Take weak limit points x_1, x_2 of $\pi(R(e_i))$, $\pi(L(e_i))$ which exist by boundedness, are unique by nondegenerateness and are equal because $\pi(x)x_1\pi(y) = \pi(R(x)y) = \pi(xL(y)) = \pi(x)x_2\pi(y)$ for $x, y \in B$ [Ped 79 3.12.3]. □

Lemma 4.3.4 *Let $(\mathbb{R}, \alpha, B)$ be a Banach-$*$-algebra dynamical system with B having a bounded approximate unit (e_i). Then $L^1(\mathbb{R}, B)$ has a bounded approximate unit of the form $(\varepsilon_i \otimes e_i)$ with (ε_i) an approximate unit in $L^1(\mathbb{R})$. There is a 1-1-correspondence between nondegenerate representations $\rho : L^1(\mathbb{R}, B) \to \mathcal{B}(H)$ and covariant representations (π, U) with $\pi : B \to \mathcal{B}(H)$ nondegenerate and $U : \mathbb{R} \to \mathcal{B}(H)$ strongly continuous s.t. $\pi(\alpha_t(b)) = U(t)\pi(b)U(t)^*$ $\forall t \in \mathbb{R}, b \in B$.*

Proof: Let ε_i be a net of positive functions $\varepsilon_i : \mathbb{R} \to \mathbb{R}_+$ s.t. $supp(\varepsilon_i) \subseteq [-\delta_i, \delta_i]$, $\delta_i \to 0$ and $\int dt\, \varepsilon_i(t) = 1$ for each i. Then $\int dt\, \varepsilon_i(t)\alpha_t(b) \to b$ for $b \in B$. If $b_i \to b$, then $\|e_ib_i - b\| \leq \|e_i(b_i - b)\| + \|e_ib - b\|$. Let $x \in C_c(\mathbb{R}, B)$. Then $\lim_i((\varepsilon_i \otimes e_i)x)(t) = e_i \int ds\, \varepsilon_i(s)\alpha_s(x(t-s)) \to x(t)$, i.e. $(\varepsilon_i \otimes e_i)x \to x$ and by the same argument also $x(\varepsilon_i \otimes e_i) \to x$. For the rest of the proof we can follow [Ped 79 7.6.2 to 7.6.5] verbatim using Lemma 4.3.3. We only indicate the steps:

(i) For each $x \in B^\sim$ and $\mu \in M(\mathbb{R})$ the measure algebra of $\mathbb{R}$ define:

$$(L(x, \mu)y)(t) = x \int \alpha_s(y(t - s))d\mu(s)$$

$$(R(x, \mu)y)(t) = \int y(t - s)\alpha_{t-s}(x)d\mu(s)$$

for $y \in C_c(\mathbb{R}, B)$ which extends to a double centralizer of $L^1(\mathbb{R}, B)$.

(ii) Each covariant representation (π, U) defines a nondegenerate representation $\pi \times U$ of $L^1(\mathbb{R}, B)$.

(iii) Each nondegenerate representation ρ of $L^1(\mathbb{R}, B)$ defines a covariant representation by

$$\pi_\rho(x) = \lim \rho(L(x, \delta_0)(\varepsilon_i \otimes e_i)) = \bar{\rho}(L(x, \delta_0))$$

$$U_t^\rho = \lim \rho(L(\mathbf{1}, \delta_t)(\varepsilon_i \otimes e_i)) = \bar{\rho}(L(\mathbf{1}, \delta_t))$$

(iv) $\pi_\rho \times U^\rho = \rho$ and $(\pi_{\pi \times U}, U^{\pi \times U}) = (\pi, U)$.

$\square$

Corollary 4.3.5 *The enveloping C^*-algebra of $L^1(\mathbb{R}, L^1(T_n))$ is isomorphic to $\mathbb{R} \rtimes \mathcal{A}_n$.*

<u>Proof:</u> In the regular representation $\pi_0 : L^1(T_n) \to \mathcal{B}(\mathcal{F}^s(L^2(\mathbb{R}, \mathbb{C}^n)))$ we have $U(r)\pi_0(K)U(r)^* = \pi_0(K^{\alpha_r})$ for $K \in L^1(T_n)$. Now π_0 is faithful and $\alpha_r = Ad(U(r))$ for the action α defined on $\mathcal{A}_n$. Thus the α's defined here and before are the same. By the foregoing Lemma, $(\mathbb{R}, \alpha, L^1(T_n))$ and $(\mathbb{R}, \alpha, \mathcal{A}_n)$ have the same covariant representations, and the representations of the corresponding crossed products may be identified. $\square$

4.3.2 Morita Equivalence between $\mathbb{R} \rtimes \mathcal{A}_n$ and $C^*(E_n)$

We rewrite the Banach algebra crossed product $L^1(\mathbb{R}, L^1(T_n))$ as an algebra of families of compact operators $L^1(K_n)$ over $\mathbb{R} \times \mathbb{R}$ which is a natural dilation of $L^1(K_{E_n})$. Throughout we let $\mathcal{H} = \mathcal{F}^s(L^2(\mathbb{R}, \mathbb{C}^n))$ and mean Borel measurable if we say measurable.

Let $L^1(K_n)$ be the Banach space of families of compact operators $K(x, y) : \mathcal{H}_y \to \mathcal{H}_x$ (mod $dxdy$) which are measurable, i.e. the maps

$$\mathbb{R} \times \mathbb{R} \mapsto \langle \xi, U(x)^* K(x, y) U(y) \eta \rangle$$

ARVESON'S SPECTRAL C*-ALGEBRAS

$\xi, \eta \in \mathcal{H}_0 = \mathcal{F}^s(L^2((-\infty, 0), \mathbb{C}^n))$ are measurable and $\int \int dx dy \, \|K(x,y)\| < \infty$, equiped with the L^1-norm. The product of $K, F \in L^1(K_n)$ is defined by

$$(KF)(x,y) := \int_{\mathbb{R}} ds \int_0^\infty d\lambda \, \big[K(x,y)(F(s-\lambda, y-\lambda) \otimes \mathbf{1}_\lambda) \\ + (K(x-\lambda, s-\lambda) \otimes \mathbf{1}_\lambda) F(s,y)\big]$$

and the involution by $K^*(x,y) = K(y,x)^*$.

As with $L^1(K_{E_n})$ it turns out that $L^1(K_n)$ is a Banach-$*$-algebra with bounded approximate units. Rather than showing that, we establish an isometric isomorphy between $L^1(K_n)$ and $L^1(\mathbb{R}, L^1(T_n))$.

Proposition 4.3.6 *Let $(A(x,y))_{x,y \in \mathbb{R}}$ be a family of bounded operators $A(x,y) \in \mathcal{B}(\mathcal{H}_y, \mathcal{H}_x)$. Then the following conditions are equivalent:*

(i) A is measurable.

(ii) The maps $A(x,y) U(y-x) : \mathcal{H}_x \to \mathcal{H}_x$ have the property that $(x,y) \mapsto \langle \xi, ((A(x,y)U(y-x)) \otimes \mathbf{1}_{(x,\infty)}) \eta \rangle$ is measurable for $\xi, \eta \in \mathcal{H}$.

Proof: (i) $\Rightarrow$ (ii): A measurable means $(x,y) \mapsto \langle \xi_0, U(x)^* A(x,y) U(y) \eta_0 \rangle$ measurable for $\xi_0, \eta_0 \in \mathcal{H}_0 = \mathcal{F}^s(L^2((-\infty, 0), \mathbb{C}^n))$ which is equivalent to $(x,y) \mapsto \langle \xi, ([U(x)^* A(x,y) U(y)] \otimes \mathbf{1}_{(0,\infty)}) \eta \rangle$ being measurable for $\xi, \eta \in \mathcal{H}$. But $\langle \eta, [U(t)[U(x)^* A(x,y) U(y) \otimes \mathbf{1}_{(0,\infty)}] U(t)^*] \xi \rangle$ is measurable in $x, y, t \in \mathbb{R}$ because $U(t)$ is strongly continuous. Putting $t = x$, we obtain condition (ii).

(ii) $\Rightarrow$ (i): The same argument starting from condition (ii). $\square$

Let $x \in L^1(\mathbb{R}, L^1(T_n))$ be given. Then x is a two parameter family $(K_t^{(s)})_{s,t \in \mathbb{R}}$ s.t. $s \mapsto (t \mapsto K_t^{(s)} \in \mathcal{K}(\mathcal{H}_t))$ is measurable w.r.t. the L^1-norm and $\int_{\mathbb{R}} ds (\int_{\mathbb{R}} dt \, \|K_t^{(s)}\|) < \infty$. For ξ, η in $\mathcal{H}$ and $f \in L^\infty(\mathbb{R})$, the functional $K \mapsto \int_{\mathbb{R}} dt \, f(t) \langle \xi, (K_t^{(s)} \otimes \mathbf{1}) \eta \rangle$ is bounded w.r.t. the L^1-norm. Hence $t \mapsto \frac{1}{\varepsilon} \int_t^{t+\varepsilon} dx \langle \xi, (K_x^{(s)} \otimes \mathbf{1}) \eta \rangle$ is dt-measurable for almost all $s \in \mathbb{R}$ and by letting $\varepsilon \to 0$

$$(s,t) \mapsto \langle \xi, (K_t^{(s)} \otimes \mathbf{1}) \eta \rangle$$

is $dtds$-measurable. If we put

$$K(x,y) := K_x^{(x-y)}U(x-y)$$

$U(x-y) : \mathcal{H}_y \to \mathcal{H}_x$, then it follows from Prop.4.3.6 and the definition of measurability of the families in $L^1(T_n)$ that this defines a family in $L^1(K_n)$. Conversely, for $K \in L^1(K_n)$ put

$$K_t^{(s)} := K(t, t-s)U(s)^* : \mathcal{H}_t \to \mathcal{H}_t$$

Remark 4.3.7 *For $K_x \in \mathcal{K}(\mathcal{H}_x)$, $x < t$ we have the identity*

$$(K_x \otimes \mathbf{1}_{(x,t)})U(r) = K(x, x-r) \otimes \mathbf{1}_{t-x}$$

where on the l.h.s. we have the composition of $U(r) : \mathcal{H}_{t-r} \to \mathcal{H}_t$ and $K_x \otimes \mathbf{1}_{(x,t)} : \mathcal{H}_x \otimes \mathcal{H}_{(x,t)} \to \mathcal{H}_x \otimes \mathcal{H}_{(x,t)}$ and on the r.h.s. the tensor product of $K(x, x-r) = K_x U(r) : \mathcal{H}_{x-r} \to \mathcal{H}_x$ and $\mathbf{1}_{t-x} : \mathcal{H}_{(0,t-s)} \to \mathcal{H}_{(0,t-s)}$ using $\mathcal{H}_{(0,a)} \cong \mathcal{H}_{(x,x+a)}$ for any $a > 0$.

Proof: This may be checked on exp-vectors. □

Proposition 4.3.8 *The above correspondence defines isomorphisms of Banach-$*$-algebras.*

Proof: We only have to check that the product is preserved: Let $k, l \in L^1(\mathbb{R}, L^1(T_n))$ where $k(r) = (K_t^{(r)})$ and $l(s) = (L_t^{(s)})$. Put

$$(k*l)(s) = \int_\mathbb{R} dr \, k(r)\alpha_r(l(s-r)) =: (R_t^{(s)})$$

In order to obtain $R_t^{(s)}$, we have to compute the product of $k(r)$ and $\alpha_r(l(s-r))$ for each r and integrate over r:

$$R_t^{(s)} = \int_\mathbb{R} dr \int_{-\infty}^t dx \big[(K_x^{(r)} \otimes \mathbf{1})(L^{(s-r)})_t^{\alpha_r} + K_t^{(r)}((L^{(s-r)})_x^{\alpha_r} \otimes \mathbf{1})\big]$$

where by L^{α_r} we mean the family in $L^1(T_n)$ as before Prop.4.3.2. $L^{(s-t)}$ is such a family and $(L^{(s-t)})^{\alpha_r}$ its image under α_r. Finally $(L^{(s-t)})_t^{\alpha_r}$ means the value of the section at t. By definition of α_r, we have $(L^{(s-r)})_t^{\alpha_r} = U(r)L_{t-r}^{(s-r)}U(r)^*$. Using Rem.4.3.1

$$(L^{(s-r)})_x^{\alpha_r} \otimes \mathbf{1}_{t-x} = U(r)(L_{x-r}^{s-r} \otimes \mathbf{1}_{t-x})U(r)^* \quad [8]$$

Hence

$$\begin{aligned}R_t^{(s)} &= \int_{\mathbb{R}} dr \int_{-\infty}^{t} dx \left[(K_x^{(r)} \otimes \mathbf{1}_{t-x})U(r)L_{t-r}^{(s-r)}U(r)^* + \right.\\ &\qquad\qquad\qquad\left. + K_t^{(r)}U(r)(L_{x-r}^{(s-r)} \otimes \mathbf{1}_{t-x})U(r)^* \right]\\ &= \int_{\mathbb{R}} dr \int_{-\infty}^{t} dx \left[(K_x^{(r)} \otimes \mathbf{1}_{t-x})U(r)L_{t-r}^{(s-r)}U(s-r) + \right.\\ &\qquad\qquad\qquad\left. + K_t^{(r)}U(r)(L_{x-r}^{(s-r)} \otimes \mathbf{1}_{t-x})U(s-r) \right]U(s)^*\end{aligned}$$

or by Rem.4.3.7

$$R_t^{(s)}U(s) = R(t, t-s) =$$

$$= \int_{\mathbb{R}} dr \int_{-\infty}^{t} dx \left[(K(x, x-r) \otimes \mathbf{1}_{t-x})L(t-r, t-s) + K(t, t-r)(L(x-r, x-s) \otimes \mathbf{1}_{t-x}) \right]$$

as a map from $\mathcal{H}_{t-s} \to \mathcal{H}_t$. Substituting $t - x = \lambda$ and $t - r = u$, we get

$$R(t, t-s) = \int_{\mathbb{R}} du \int_0^{\infty} d\lambda \left[(K(t-\lambda, u-\lambda) \otimes \mathbf{1}_\lambda)L(u, t-s) + K(t, u)(L(u-\lambda, t-s-\lambda) \otimes \mathbf{1}_\lambda) \right]$$

which concludes the proof. □

Thus $\mathbb{R} \rtimes \mathcal{A}_n$ is the enveloping C^*-algebra $C^*(L^1(K_n))$ of $L^1(K_n)$.

Consider the subset of $L^1(K_n)$ defined by $L := \{K \in L^1(K_n) |\ K(x, y) = 0$ for $y \leq 0$ and $K(x, y) \in \mathcal{K}(\Omega_0 \otimes \mathcal{H}_{(0,y)}, \mathcal{H}_x)$ for $y > 0\}$ where $\Omega_0 = exp(0|(-\infty, 0))$ is the vacuum in $\mathcal{F}^s(L^2((-\infty, 0), \mathbb{C}^n))$. Let $B := \{K \in L^1(K_n) |\ K(x, y) = 0$ if $x \leq 0$ or $y \leq 0$ and $K(x, y) \in \mathcal{K}(\Omega_0 \otimes \mathcal{H}_{(0,y)}, \Omega_0 \otimes \mathcal{H}_{(0,x)})$ for $x, y > 0\}$.

---
[8] As before the index at $\mathbf{1}$ denotes an interval length

Lemma 4.3.9 *(i) L is a left ideal and B is a subalgebra of $L^1(K_n)$ isomorphic to $L^1(K_{E_n})$ and equal to L^*L.*

(ii) $\overline{\bigcup_{t \in \mathbb{R}} \alpha_t(B)} = L^1(K_n)$.

(iii) $LL^* = L^1(K_n)$.

(iv) Each nondegenerate representation π of $L^1(K_{E_n})$ induces a representation $\bar{\pi}$ of $L^1(K_n)$ s.t. $\|\bar{\pi}(x)\| \geq \|\pi(x)\|$ for each $x \in B$. In particular, the closure of B in $C^*(L^1(K_n))$ is equal to $C^*(E_n)$.

Proof: (i): It is immediate that L is a left ideal. Using $\mathcal{H}_{(0,x)} = E_n(x)$, $\mathcal{H}_{(0,y)} = E_n(y)$, we have $\mathcal{K}(\Omega_0 \otimes \mathcal{H}_{(0,y)}, \Omega_0 \otimes \mathcal{H}_{(0,x)}) = \Theta_{\Omega_0, \Omega_0} \otimes \mathcal{K}(E_n(y), E_n(x))$. One can see from the definition of the multiplication that B is isomorphic to $L^1(K_{E_n})$ under this identification. Moreover, $B \subseteq L$ and hence $BB \subseteq L^*L$ and we also have $L^*L \subseteq B$. But $BB = B$ because B contains approximate units ([HeRo II (32.26)]).

(ii): Define $\hat{E}_n := \bigcup_{t \in \mathbb{R}} \{t\} \times \mathcal{F}^s(L^2((-\infty,t),\mathbb{C}^n)) \subseteq \mathbb{R} \times \mathcal{F}(L^2(\mathbb{R},\mathbb{C}^n))$ considered as a Borel subspace. $\hat{E}_n$ is a trivial Borel fibration over $\mathbb{R}$ and $L^1(K_n)$ may be identified with L^1-sections in a trivial Borel fibration over $\mathbb{R} \times \mathbb{R}$. The subsets $E_n^c := \bigcup_{x>c} \{x\} \times \Omega_c \otimes \mathcal{H}_{(c,x)}$ are equal to E_n as Borel fibrations over $\mathbb{R}_+$ translated by c. Hence we find trivializing sections $e_n^c(t)$ with $t > c$ in E_n^c and the following sections in $L^1(K_n)$:

$$e_{nm}^c(x,y) = \begin{cases} e^{2c-x-y}\Theta_{\Omega_c \otimes e_n^c(x), \Omega_c \otimes e_m^c(y)} & \text{if } x,y > c \\ 0 & \text{otherwise} \end{cases}$$

If (c_k) is a sequence of real numbers $c_k \to -\infty$, then the set $\{e_{nm}^{c_k}|n,m,k \in \mathbb{N}\}$ is generating in the sense of Appendix Prop.7 because $\overline{\bigcup_{s<t} \Omega_s \otimes \mathcal{H}_{(s,t)}} = \mathcal{H}_t$. If we now define $B_s = \{K \in L^1(K_n) | K(x,y) = 0 \text{ if } x \leq s \text{ or } y \leq s \text{ and } K(x,y) \in \mathcal{K}(\Omega_s \otimes \mathcal{H}_{(s,y)}, \Omega_s \otimes \mathcal{H}_{(s,x)}) \text{ otherwise }\}$, then $B_s \cong L^1(K_{E_n})$ and $B_s = U(s)BU(s)^* = \alpha_s(B)$. Therefore $\bigcup_{s \in \mathbb{R}} B_s$ contains a generating sequence.

(iii): $B \subseteq LL^*$ and from the multiplication formula $U(t)LL^*U(t)^* =$

$\alpha_t(LL^*) = LL^*$ for any $t \in \mathbb{R}$. The claim is now clear by (ii).

(iv): Suppose $\pi : L^1(K_{E_n}) \to \mathcal{B}(H_\pi)$ is a nondegenerate representation. Then the algebraic tensor product $L \odot_B H_\pi$ over B is a prehilbert space with the scalar product $\langle x \otimes \xi, y \otimes \eta \rangle = \langle \xi, \pi(x^*y)\eta \rangle$ for $x, y \in L$ and $\xi, \eta \in H_\pi$. There is the action $\bar{\pi}(z)(x \otimes \xi) = zx \otimes \xi$ of $L^1(K_n)$ which is bounded because π is bounded. For (u_i) an approximate unit of $L^1(K_{E_n})$ and $z \in B$ we have:

$$\langle u_i \otimes \xi, \bar{\pi}(z)(u_i \otimes \eta) \rangle \to \langle \xi, \pi(z)\eta \rangle$$

which implies $\|\bar{\pi}(z)\| \geq \|\pi(z)\|$. □

Theorem 4.3.10 $C^*(E_n)$ is (strongly) Morita equivalent to $\mathbb{R} \rtimes \mathcal{A}_n$ for each $n \in \mathbb{N}$. $\mathbb{R} \rtimes \mathcal{A}_n$ is the inductive limit of copies of $C^*(E_n)$ and $C^*(E_n)$ is KK-contractible.

<u>Proof:</u> Let $X := \overline{L}$ be the closure of L in the universal representation of $\mathbb{R} \rtimes \mathcal{A}_n$. By Lemma 4.3.9.(iv), the closure $\overline{B}$ of $B = L^*L$ in $C^*(L^1(K_n)) = \mathbb{R} \rtimes \mathcal{A}_n$ is equal to $C^*(E_n)$. We have $L^*L \subseteq X^*X \subseteq \overline{L^*L} = \overline{B} \subseteq X^*X$ and $XX^* = \mathbb{R} \rtimes \mathcal{A}_n$ by part (iii). Therefore $\mathbb{R} \rtimes \mathcal{A}_n$ and $C^*(E_n)$ are (strongly) Morita equivalent with X being the full Morita equivalence module (c.f.[Rie 82]). We have $KK(\mathcal{A}_n, \mathcal{A}_n) = 0$ (c.f.Cor.3.4.2) for each $n \in \mathbb{N}$ and hence by [FaSk 81 Thm.1], $KK(\mathbb{R} \rtimes \mathcal{A}_n, \mathbb{R} \rtimes \mathcal{A}_n) = 0$. Thus $C^*(E_n)$ and 0 are KK-equivalent using Morita invariance of KK-theory and the Kasparov product as in Cor.3.4.2. The second assertion follows from part (i) and (ii) of Lemma 4.3.9. □

Remark 4.3.11 (i) There is also a stable version of the subalgebra $C^*(E_n) \subseteq \mathbb{R} \rtimes \mathcal{A}_n$ which is the closure of $\{K \in L^1(K_n) | K(x,y) = 0 \text{ if } x < 0 \text{ or } y < 0\}$ using $\mathcal{K}(\mathcal{H}_y, \mathcal{H}_x) = \mathcal{K} \otimes \mathcal{K}(E_n(y), E_n(x))$. It is equal to $\mathcal{K} \otimes C^*(E_n)$.

(ii) With the same argument as in Prop.4.3.9 we find subalgebras $B_v = \{K \in L^1(K_E) | K(x,y) = 0 \text{ if } x \leq t \text{ or } y \leq t \text{ and } K(x,y) \in \mathcal{K}(vE(x -$

$t), vE(y-t))$ otherwise $\}$ of $L^1(K_E)$, E any product system and $v \in E(t)$ any unit vector. The closure of B_v is equal to $C^*(E)$ and it is the image of an endomorphism φ_v of $C^*(E)$ s.t. $\varphi_v(f \otimes \bar{g}) = vf \otimes \overline{vg}$. If E contains a unit, then this endomorphism is inner.

4.3.3 Simplicity

We give a version of Arveson's proof of simplicity for our dilated algebra.
There are the following representations of $\mathbb{R} \rtimes \mathcal{A}_n$:

(i) The regular representation ρ induced by $\pi_0 : \mathcal{A}_n \to \mathcal{B}(\mathcal{H})$, $\mathcal{H} = \mathcal{F}^s(L^2(\mathbb{R}, \mathbb{C}^n))$ acting on $L^2(\mathbb{R}, \mathcal{H})$ which is faithful because π_0 is faithful.

(ii) The shift representation $\sigma : \mathbb{R} \rtimes \mathcal{A}_n \to \mathcal{B}(\mathcal{H})$ given by the covariant representation (π_0, U) where $U(t) = exp(S_t)$.

σ acts as follows:

$$\sigma(K)exp(f) = \int_{\mathbb{R}} dx \int_{\mathbb{R}} dy \, [K(x,y)(exp(f_y))] \otimes exp(S_{x-y}f^y)$$

$f \in L^2(\mathbb{R}, \mathbb{C}^n)$ for $K \in L^1(K_n)$. Let $\sigma_s = \sigma|C^*(E_n)_s$ where $C^*(E_n)_s$ is the closure of $B_s = \alpha_s(B) \subseteq L^1(K_n)$ as in Lemma 4.3.9.(ii). Then σ_s is degenerate because $\overline{\sigma(B_s)\mathcal{H}} = \Omega_s \otimes \mathcal{H}^s$. Note that σ (σ_s) is the GNS-representation associated to $\omega = \langle \Omega, \cdot \, \Omega \rangle$ ($\omega_s = \langle \Omega^s, \cdot \, \Omega^s \rangle$, $\Omega^s = exp(0|[s, \infty))$)

ρ acts as follows:

$$\rho(K)(\varphi \otimes exp(f)) = \int_{\mathbb{R}} dx \int_{\mathbb{R}}^{\oplus} dy \, (S_{x-y}\varphi) \otimes [[K(x,y)exp(f_y)] \otimes exp(S_{x-y}f^y)]$$

$\varphi \in L^2(\mathbb{R})$, $f \in L^2(\mathbb{R}, \mathbb{C}^n)$. [9]

Let γ be the gauge group on $\mathbb{R} \rtimes \mathcal{A}_n$ i.e. the dual action of α. Then we can form the representation $\bar{\sigma} := \int_{\mathbb{R}}^{\oplus} \sigma \circ \gamma_{-t} \, dt$ which acts on $L^2(\mathbb{R}, \mathcal{H})$.

Lemma 4.3.12 $\bar{\sigma}$ and ρ are unitarily equivalent

---
[9] the first tensor product means the tensor product in $L^2(\mathbb{R}) \otimes \mathcal{H}$ whereas the second is a tensorproduct inside $\mathcal{H}$

ARVESON'S SPECTRAL C*-ALGEBRAS

<u>Proof:</u> Let $F : L^2(\mathbb{R}, \mathcal{H}) \to L^2(\mathbb{R}, \mathcal{H})$ be the Fourier transform

$$(F\xi)(s) = \tfrac{1}{\sqrt{2\pi}} \int_\mathbb{R} dt \; e^{ist}\xi(t)$$

($\xi \in L^1 \cap L^2$). Then for $K \in L^1(K_n)$ and $\varphi \in L^1 \cap L^2$

$(F(\bar\sigma(K)(\varphi \otimes exp(f))))(t) =$

$$= \tfrac{1}{\sqrt{2\pi}} \int_\mathbb{R} dx \int_\mathbb{R} dy \int_\mathbb{R} ds \; e^{ist}e^{-is(x-y)}\varphi(s)[K(x,y)exp(f_y)] \otimes exp(S_{x-y}f^y)$$

$$= \tfrac{1}{\sqrt{2\pi}} \int_\mathbb{R} dx \int_\mathbb{R} dy \int_\mathbb{R} ds \; e^{is(t-x+y)} \varphi(s) [K(x,y)exp(f_y)] \otimes exp(S_{x-y}f^y)$$

On the other hand

$(\rho(K)(F(\varphi \otimes exp(f))))(t) =$

$$= \tfrac{1}{\sqrt{2\pi}} \int_\mathbb{R} dx \int_\mathbb{R} dy \int_\mathbb{R} ds \; e^{is(t-x+y)}\varphi(s)[K(x,y)exp(f_y)] \otimes exp(S_{x-y}f^y)$$

which is exactly the same. □

The following Proposition is due to Arveson [Ar 90a Thm.8.2] adapted to our situation and is one of the two main parts of his proof that $C^*(E)$ is simple if E contains a unit. The other part (minimality of the regular representation λ [Ar 90a sec.7]) is not needed here. Instead we only use Lemma 4.3.12 and that ρ is faithful.

Proposition 4.3.13 *For each representation π of $\mathbb{R} \rtimes \mathcal{A}_n$, $x \in \mathbb{R} \rtimes \mathcal{A}_n$ and $t \in \mathbb{R}$*

$$\|\sigma \circ \gamma_t(x)\| \leq \|\pi(x)\|$$

<u>Proof:</u> By Thm.4.3.10, we may assume $x \in C^*(E_n)_s := \alpha_s(C^*(E_n))$ for some $s \in \mathbb{R}$. Define $\omega^t := \omega \circ \gamma_t$ with γ the dual action to α. Observe

that $\gamma_t(C^*(E_n)_s) = C^*(E_n)_s$ and $(\sigma \circ \gamma_t)|C^*(E_n)_s = \sigma_s \circ \gamma_t$. Using $C^*(E_n)_s \cong C^*(E_n)$, it follows from [Ar 90a Thm.8.2] that there is a sequence of unit vectors (ξ_n) in the representation space H of π s.t. $\omega^t(x) = \lim_{n\to\infty}\langle \xi_n, \pi(x)\xi_n \rangle$. Hence $\|\pi(x)\| \geq \|\sigma_s \circ \gamma_t(x)\|$ for each $t \in \mathbb{R}$. $\square$

Corollary 4.3.14 $\mathbb{R} \rtimes \mathcal{A}_n$ *is simple.*

<u>Proof:</u> $\|\rho(x)\| \stackrel{4.3.12}{=} \|\bar{\sigma}(x)\| \stackrel{4.3.13}{=} \|\sigma(x)\| \stackrel{4.3.13}{\leq} \|\pi(x)\| \leq \|\rho(x)\|$ for any representation π of $\mathbb{R} \rtimes \mathcal{A}_n$ where the last inequality comes from the faithfulness of ρ [Ped 79 7.7.7]. $\square$

Corollary 4.3.15 $C^*(E_n) = C^*_r(E_n)$.

<u>Proof:</u> By [Br 77], $C^*(E_n)$ and $\mathbb{R} \rtimes \mathcal{A}_n$ are stably isomorphic. Hence $C^*(E_n)$ is simple (because A is simple iff $A \otimes \mathcal{K}$ is simple for any C^*-algebra). $\square$

4.3.4 Infiniteness

Let A be a C^*-algebra and $p, q \in A$ projections. p and q are called Murray-von Neumann equivalent if there is a partial isometry $s \in A$ s.t. $s^*s = p$ and $ss^* = q$. A projection p is called infinite if it is equivalent to a proper subprojection of itself. A is called finite if it does not contain any infinite projections, and it is called stably finite if the same holds for each finite matrix algebra $M_k(A)$ over A. By the work of Cuntz, Blackadar, Haagerup and Kirchberg, it is known that for A simple nuclear and separable A is stably finite iff there is a lower semicontinuous trace on $M_\infty(\mathcal{P}(A))$, $\mathcal{P}(A)$ the Pedersen ideal of A. We call A infinite if it is not stably finite.

Proposition 4.3.16 *Let A be a simple C^*-algebra containing a nontrivial projection p (i.e. $p \neq 0$ and $p \neq 1$ if A has a unit) s.t. $K_0(A) = 0$. Then A is infinite.*

ARVESON'S SPECTRAL C^*-ALGEBRAS

<u>Proof:</u> If A is unital, then by the definition of K-theory $[p] = [p \oplus p]$. Hence in some matrix algebra $M_k(A)$ we find a partial isometry s s.t.

$$s^*s = \begin{pmatrix} p & & \\ & 0 & \\ & & q \end{pmatrix} \quad \text{and} \quad ss^* = \begin{pmatrix} p & & \\ & p & \\ & & q \end{pmatrix}$$

where q is a projection in $M_{k-2}(A)$. Hence $p \oplus p \oplus q$ is infinite.

If A is not unital, then $pAp \subseteq A$ as well as

$$M_2(pAp) = \begin{pmatrix} p & \\ & p \end{pmatrix} M_2(A) \begin{pmatrix} p & \\ & p \end{pmatrix} \subseteq M_2(A)$$

are full hereditary unital C^*-subalgebras and thus by [Br 77] are all stably isomorphic to A. $M_2(pAp)$ contains nontrivial projections and the first case applies. $\square$

Corollary 4.3.17 $C^*(E_n)$ *is infinite for each* $n \in \mathbb{N}$.

Summarizing we have:

Theorem 4.3.18 *The spectral C^*-algebras $C^*(E_n)$ for the exponential product systems E_n are infinite simple unitless nuclear and KK-contractible C^*-algebras (In particular they fulfil UCT).*

Appendix

A. Bochner Integrals

Let (Ω, μ) be a σ-finite measure space and X a Banach space. $f : \Omega \to X$ of the form $\sum_{i=1}^{N} \chi_{\Delta_i} x_i$ with $x_i \in X$ and χ_{Δ_i} the characteristic functions of μ-measurable subsets s.t. $\mu(\Delta_i) < \infty$ is called simple. $f : \Omega \to X$ any function is called strongly μ-measurable if there exists a sequence of simple functions $(f_n)_{n \in \mathbb{N}}$ s.t. $f_n(t) \to f(t)$ μ-a.e. in norm. f is called weakly μ-measurable if for each $x^* \in X^*$ the function $t \mapsto x^*(f(t))$ is μ-measurable.

Theorem 1 *(Pettis) $f : \Omega \to X$ is strongly μ-measurable iff f is weakly μ-measurable and there exists a zero set N s.t. $f(\Omega \setminus N)$ is norm separable.*

Proof: [Die 84 p.25]. □

f is called Bochner integrable if there is a sequence (f_n) of simple functions $f_n = \sum_i \chi_{\Delta_i^n} x_i^n$ s.t. $\lim_{n \to \infty} \int d\mu(t) \, \|f_n(t) - f(t)\| = 0$ and $\lim_n \sum_i \mu(\Delta_i^n) x_i^n$ exists. In this case $\int d\mu \, f$ is defined as this limit.

Theorem 2 *(i) $f : \Omega \to X$ μ-measurable is Bochner integrable iff $\int_\Omega d\mu \, \|f\| < \infty$.*

(ii) If (f_n) are Bochner integrable s.t. f is the pointwise limit of (f_n) μ-a.e. and there is a μ-integrable g s.t. $\|f_n(t)\| \leq g(t)$, then f is Bochner integrable, $\int d\mu \, \|f - f_n\| \to 0$ and $\int d\mu \, f_n \to \int d\mu \, f$.

(iii) $\|\int_\Delta d\mu \, f\| \leq \int_\Delta d\mu \, \|f\|$ for f Bochner integrable and $\Delta \subseteq \Omega$ μ-measurable.

APPENDIX

Proof: [Die 84 p.26-27] □

The Banach space of all Bochner integrable functions (mod μ) is denoted by $L^1(\Omega, \mu; X)$.

The following remark is sometimes used implicitely.

Remark 3 *Let $\mathcal{K}$ be the compact operators on a separable Hilbert space H. Then $K : \Omega \to \mathcal{K}$ is Bochner integrable if $t \mapsto \|K(t)\| \in L^1(\Omega, \mu)$ and $t \mapsto \langle \xi, K(t)\eta \rangle$ is μ-measurable for all $\xi, \eta \in H$. Furthermore, $\xi \mapsto \int_\Omega d\mu(t) K(t)\xi$ is a compact operator.*

Proof: Because countable sums of measurable functions are measurable the assumption on K implies $t \mapsto \operatorname{tr}(TK(t))$ is measurable for each trace class operator T. Thus $t \mapsto K(t)$ is norm measurable by Thm.1. □

Remark 4 *Let $A : E \to F$ be a bounded linear operator between Banach spaces and $f : \Omega \to E$ μ-Bochner integrable. Then $t \mapsto A(f(t))$ is μ-Bochner integrable and $\int d\mu(t)\, A(f(t)) = A(\int d\mu(t)\, f(t))$.*

Recall that a regular Borel measure on a locally compact space is a measure μ defined on the Borel sets which is finite on compact subsets s.t. $\mu(\Lambda) = \sup_{K \subseteq \Lambda} \mu(K) = \inf_{\Lambda \subseteq O} \mu(O)$ where K, O run through the compact, open subsets of Γ respectively. On a second countable locally compact space the positive regular Borel measures are given by positive linear forms on $C_c(\Gamma)$. Two measures are called equivalent if they have the same zero sets. The corresponding equivalence classes are denoted by $[\mu]$.

Remark 5 *(i) Let μ be a positive regular Borel measure on $\mathbb{R}_+$ quasi invariant under translations by $t \in \mathbb{R}_+$ (i.e. μ and $\mu(\cdot + t)$ are equivalent measures on $\mathbb{R}_+$ for each $t > 0$). Then $[\mu] = [dt]$.*

(ii) Let μ be a positive regular Borel measure on the interval $(0, 1)$ and $\mu_t := \mu|(0, t)$. Suppose $[\mu] = [\mu_t + T_t \mu_{1-t}]$ for each $t \in (0, 1)$ where $T_t \mu = \mu(\cdot - t)$ is the translation to the right. Then $[\mu] = [dt]$.

Proof: (i): By translating $\mu|(0,1)$ to the intervals $(-1, 0), (-2, -1), \ldots$ and taking the sum, we obtain a measure on $\mathbb{R}$ quasi invariant under positive hence also under negative translations.

(ii): Extending μ as in (i) to both sides, we obtain a positive regular Borel measure λ on $\mathbb{R}$ quasi invariant under translations. □

Remark 6 *For μ a regular Borel measure on an open interval I and $f \in L^1(I, \mu; X)$ with X a Banach space, we have*

$$f(t) = \lim_{\varepsilon \to 0} \tfrac{1}{\mu([t,t+\varepsilon))} \int_t^{t+\varepsilon} d\mu(x)\, f(x)$$

μ-a.e.

Proof: For $X = \mathbb{C}$ this is the Radon-Nikodym Theorem. Now proceed as in [DiUh 77 p.49 Thm.9]. □

Following the theory of direct integrals, consider a Borel fibration $q : V \to \Gamma$ where Γ is a second countable locally compact space s.t. $V(t) := q^{-1}(t)$ is a separable Banach space and for some $t_0 \in \Gamma$ there is a Borel map Ψ making

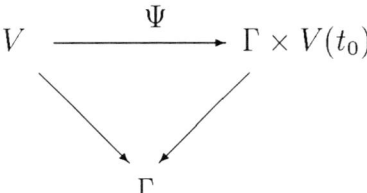

commutative.

Then we have an identification between the L^1-sections $L^1(V)$ in V and $L^1(\Gamma, V(t_0))$. Suppose μ is a regular Borel measure on Γ. We call a sequence $(f_n) \subseteq L^1(V)$ generating (w.r.t. the measure μ) if $\overline{\mathrm{span}_{\mathbb{C}}\{f_n(t)|n \in \mathbb{N}\}} = V(t)$ μ-a.e.

Proposition 7 *Let $X \subseteq L^1(V)$ be a linear subspace of L^1-sections of V containing a generating sequence and closed under multiplication with continuous functions of compact support. Then X is normdense.*

Proof: We have $L^1(V) \cong L^1(\Gamma, \mu; V(t_0))$ and $L^1(V)^* \cong L^\infty(\Gamma, \mu; V(t_0)^*)$. Let $(f_n) \subseteq X$ be generating, $\omega \in L^\infty(\Gamma, \mu; V(t_0)^*)$ s.t. $\langle \omega, X \rangle = 0$ and χ_Δ the characteristic function of an open set $\Delta \subseteq \Gamma$ with compact closure. Then there is a sequence of continuous functions with compact support (φ_N)

APPENDIX

s.t. $\varphi_N(t) \searrow \chi_\Delta(t)$ for all t. $\varphi_N f_n \in X$ for $N, n \in \mathbb{N}$ by assumption and $\langle \omega, \chi_\Delta f_n \rangle = \int_\Delta \langle \omega(t), f_n(t) \rangle d\mu(t) = \lim_{N \to \infty} \int d\mu \, \varphi_N(t) \langle \omega(t), f_n(t) \rangle = 0$ for each n. Thus $\Delta_n := \{t \in \Gamma | \langle \omega(t), f_n(t) \rangle \neq 0\}$ is a zero set and so is $\bigcup_n \Delta_n$ therefore $\omega(t) = 0$ a.e. □

B. Direct Integrals

B.1. Definition

Let $(\mathcal{K}_t)$ be a family of Hilbert spaces (we allow $\mathcal{K}_t = 0$) indexed by numbers in an open interval I and μ a σ-finite measure on I [10]. A subset $\mathcal{S} \subseteq \prod_{t \in I} \mathcal{K}_t$ is called admissible if the following properties are satisfied:

(i) $\mathcal{S}$ is a linear subspace s.t. $f \in \mathcal{S}$ implies $\chi_\Delta f \in \mathcal{S}$ for each $\Delta \in \mathcal{B}(I)$.

(ii) For each $f \in \mathcal{S}$ the function $t \mapsto \|f(t)\|$ is Borel measurable and $\int_I \|f(t)\|^2 d\mu(t) < \infty$.

(iii) $[\{f(t) | f \in \mathcal{S}\}] = \mathcal{K}_t$, μ a.e.

A function $f \in \prod_{t \in I} \mathcal{K}_t$ is called strongly $\mathcal{S}$-measurable if there is a sequence $(t_n) \subseteq \mathcal{S}$ s.t. $\lim_{n \to \infty} \|f(t) - t_n(t)\|_t = 0$, μ-a.e. Define on the space of strongly $\mathcal{S}$-measurable functions $f \in \prod_{t \in I} \mathcal{K}_t$ s.t. $\int \|f(t)\|^2 d\mu(t) < \infty$ the semiscalar product $\langle f, g \rangle = \int_I \langle f(t), g(t) \rangle_t d\mu(t)$. The completion of the resulting prehilbert space is called the direct integral $\int_I^\oplus \mathcal{K}_t d\mu$. We denote the image of $\mathcal{S}$ again by $\mathcal{S}$ (which is fully justified if $supp\,\mu = I$). We call $\mathcal{S}$ separable if $\int_I^\oplus \mathcal{K}_t d\mu$ is separable. For $\mathcal{S}$ separable the function $t \mapsto \dim \mathcal{K}_t$ is Borel measurable and the sets $\Lambda_n = \{t \in I | \dim \mathcal{K}_t = n\}$, $n \in \mathbb{N} \cup \{\infty\}$ are Borel measurable. We denote by $\int_{\Lambda_n}^\oplus \mathcal{K}_t d\mu$ the direct integral defined by $\mathcal{S}|\Lambda_n$. There is an action of $L^\infty(I, \mu)$ on $\int_I^\oplus \mathcal{K}_t d\mu$ defined by $(m(\alpha)f)(t) = \alpha(t)f(t)$ which is a weakly closed subalgebra, also called the scalar operators. We have $\|m(\alpha)\| = \|\alpha\|_\infty$. The commutant of $\{m(\alpha) | \alpha \in L^\infty\}$ is the von Neumann algebra of diagonal operators given by functions $A \in \prod_{t \in I} \mathcal{B}(\mathcal{K}_t)$ s.t. $A\mathcal{S} \subseteq \mathcal{S}$ and $\mathrm{essup}(\|A(t)\|) < \infty$.

---

[10] we could take here Γ as before but we don't need this generality

Proposition 8 $\int_{\Lambda_n}^{\oplus} \mathcal{K}_t d\mu \cong L^2(\Lambda_n, \mu|\Lambda_n, \mathbb{C}^n)$ by a unitary V commuting with the L^∞-multiplication.

Proof: [BaWo 83 Prop.15] □

Thus the sequence of pairwise disjoint measure classes $\{[\mu_1], \ldots, [\mu_\infty]\}$ where $\mu_i := \mu|\Lambda_i$ (extended by 0) determines the direct integral. We call two direct integrals equivalent if μ_n and ν_n are equivalent measures for each $n \in \mathbb{N}$.

We say that two direct integrals $\int_I^\oplus \mathcal{K}_t d\mu$ and $\int_{I'}^\oplus \mathcal{L}_{t'} d\nu$ differ by a change of measure given by a monotone invertible map $\tau : I' \to I$ if $\mu \circ \tau$ and ν are equivalent measures on I' and the map $V_\tau : f \mapsto \sqrt{\frac{d\mu \circ \tau}{d\nu}} f \circ \tau$ extends to a unitary transformation which may also be called a change of measure.

Remark 9 Suppose $V : \int_I^\oplus \mathcal{K}_t d\mu \to \int_{I'}^\oplus \mathcal{L}_{t'} d\nu$ is a unitary s.t. there exists $\tau : I' \to I$ s.t. $f|(I \setminus \tau(\Delta')) = 0$ implies $Vf|(I' \setminus \Delta') = 0$ for each $\Delta' \in \mathcal{B}_0(I')$ and $f \in \int_{I'}^\oplus \mathcal{L}_{t'} d\nu$. Then $V = V_\tau$ up to a diagonal unitary.

Proof: Let V be as asserted. W.l.o.g. suppose $\Lambda_0 = \emptyset = \Lambda_0'$. Then the assumption implies $\mu \circ \tau(\Delta') = 0 \Leftrightarrow \nu(\Delta') = 0$ hence $\frac{d\mu \circ \tau}{d\nu}$ exists. Suppose first $\tau(\Lambda_n') = \Lambda_n \ \forall n \in \mathbb{N}$. Using Prop.8 we may define a unitary V_τ by $f \mapsto \sqrt{\frac{d\mu \circ \tau}{d\nu}} f \circ \tau$. The map $V_\tau^{-1} \circ V$ makes the support of sections only smaller. Let π be the action of the scalar operators. Then $\chi_\Delta \chi_\Delta = \chi_\Delta$ implies $V_\tau^{-1} \circ V \pi(m_1) \pi(m_2) = \pi(m_1) V_\tau^{-1} \circ V \pi(m_2)$ provided the m_i are equal to finite sums of the form $\sum_i \alpha_i \chi_{\Delta_i}$, $\Delta_i \in \mathcal{B}_0(I)$. Thus $V_\tau^{-1} \circ V \in \pi(L^\infty)'$ and this commutant are the diagonal operators. To show $\tau(\Lambda_n') = \Lambda_n$ for $n \in \mathbb{N}$, suppose there is a Borel set $\bar{\Lambda}_n' \subseteq \Lambda_n'$ of positive measure s.t. $\tau(\bar{\Lambda}_n') = \bar{\Lambda}_m \subseteq \Lambda_m$. For $s = \sum \alpha_i \chi_{\Delta_i \cap \bar{\Lambda}_m}$, $\Delta_i \in \mathcal{B}_0(I)$ a finite sum, it follows that $V_\tau \circ \pi(s) = \pi(s \circ \tau) V_\tau$ and this relation extends to $L^\infty(\bar{\Lambda}_m)$. Thus multiplicity theory implies $m = n$. □

Remark 10 For $\mu = \nu = dt$ and $I = I'$ any such function τ is continuous with derivative in $L^1(I, dt)$.

B.2. Direct Integrals Defined by Families of Positive Definite Functions

Let S be a set and $k(t;\cdot,\cdot)$ be a family of positive definite functions on S indexed by an interval I together with a measure μ s.t. for any $x,y \in S$ the function $t \mapsto k(t;x,y)$ lies in $L^1(I,\mu)$. Define $\mathcal{K}_t$ as the kernel Hilbert space of $(S, k(t;\cdot,\cdot))$. Let $J_t : S \to \mathcal{K}_t$ be the identification map from S to $\mathcal{K}_t$ and put

$$\mathcal{S} := \Big\{ \sum_{i=1}^m \alpha_i(t) J_t s_i \,\big|\, s_i \in S \text{ and } \alpha_i \in L^\infty(I,\mu) \Big\}$$

Then $\mathcal{S} \subseteq \prod_{t \in I} \mathcal{K}_t$ is admissible and defines a direct integral. It will be separable if S is countable.

B.3. The Spectral Theorem in the Direct Integral Version

We assume H to be a fixed separable Hilbert space. By a projection we always mean a projection in H.

Definition 11 *(i) A spectral measure on $I = (a,b)$ is a projection valued measure $\Delta \mapsto E(\Delta)$ on the Borel sets $\mathcal{B}(I)$ s.t. $E(I) = 1$, $E(\emptyset) = 0$.*

(ii) A spectral resolution (on I) is a monotone decreasing family of projections $(P_t)_{t \in I}$ s.t. $t \mapsto P_t$ is left continuous.

Remark 12 *(i) $\lim_{t \searrow t_0} P_t$ and $\lim_{t \nearrow t_0} P_t$ always exist for a spectral resolution.*

(ii) $E([s,t)) := P_s - P_t$ defines a spectral measure for each spectral resolution (E is countably additive by monotonicity of $\langle \xi, P_t \xi \rangle$ for $\xi \in H$).

Theorem 13 *For each spectral resolution on I there exists a measure μ and a unitary onto a direct integral $\int_I^\oplus \mathcal{K}_t d\mu$ s.t. P_t becomes the multiplication $m(\chi_{[t,b)})$. Thus the sequence of measure classes $\{[\mu_1], [\mu_2], \ldots, [\mu_\infty]\}$ is a complete unitary invariant of the spectral resolution.*

Proof: [BaWo 83 Prop.21 and Rem.22] □

C. Conditionally Positive Definite Functions

Definition 14 *Let X be a set and $k : X \times X \to \mathbb{C}$ a function (also called kernel)*

(i) k is called positive definite if for $\lambda_1, \ldots, \lambda_n \in \mathbb{C}$ and $x_1, \ldots, x_n \in X$ $\sum_{i,j=1}^{n} \overline{\lambda_i} \lambda_j k(x_i, x_j) \geq 0$.

(ii) k is called conditionally positive definite if for $\lambda_1, \ldots, \lambda_n \in \mathbb{C}$ and $x_1, \ldots, x_n \in X$ s.t. $\sum_{i=1}^{n} \lambda_i = 0$ we have $\sum_{i,j=0}^{n} \overline{\lambda_i} \lambda_j k(x_i, x_j) \geq 0$.

For k (conditionally) positive definite the completion of the complex valued functions with finite support (s.t. $\sum_{x \in X} f(x) = 0$) is a Hilbert space called kernel Hilbert space (conditional Hilbert space). If H_k is a kernel Hilbert space, we can find a map $\lambda : X \to H_k$ unique up to unitary equivalence s.t. $k(x,y) = \langle \lambda(x), \lambda(y) \rangle$. More precisely, for any map $\mu : X \to H$ into a Hilbert space H s.t. $\langle \mu(x), \mu(y) \rangle = k(x,y)$ there is a unique isometry $\hat{\mu} : H_k \to H$ s.t.

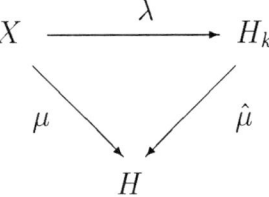

commutes.

Proposition 15 *The following conditions for a kernel are equivalent:*

(i) k is conditionally positive definite.

(ii) For any fixed $z \in X$ the kernel

$$(x,y) \mapsto k_z(x,y) := k(x,y) - k(x,z) - k(z,y) + k(z,z)$$

is positive definite.

(iii) The kernels $(x,y) \mapsto e^{-tk(x,y)}$ are positive definite for each $t > 0$.

Proof: [PaSc 72 Lemma 1.7] □

The dimension of the Hilbert spaces associated to $(x,y) \mapsto e^{-tk(x,y)}$ is the dimension of the conditional Hilbert space plus one independently of $t > 0$.

References

[Ar 89] W.Arveson : Continuous Analogues of the Fock Space, Memoirs AMS no. 409 (1989)

[Ar 90a] W.Arveson : Continuous analogues of the Fock Space II, J. Funct. Anal. 90 (1990), 138-205

[Ar 90b] W.Arveson : Continuous analogues of the Fock Space III: Singular States, J. Operator Theory 22 (1990), 165-205

[Ar 90c] W.Arveson : Continuous analogues of the Fock Space IV: Essential States, Acta Math. 164 (1990), 265-300

[Ar 90d] W.Arveson : The Spectral C^*-algebra of an E_0-semigroup, Proc. Symp. pure Math. 51 AMS (1990), 1-15

[Ar 91] W.Arveson : C^*-Algebras Associated with Sets of Semigroups of Isometries, International J. Math. 2 (1991), 235-255

[Ar 94a] W.Arveson : E_0-Semigroups in Quantum Field Theory, preprint (1994)

[Ar 94b] W.Arveson : Path Spaces, Continuous Tensor Products and E_0-Semigroups, preprint (1994)

[Ar 95] W.Arveson : Noncommutative Flows I: Dynamical Invariants, preprint (1995)

[ArWo 66] H.Araki, J.Woods : Complete Boolean Algebras of Type I Factors, Publ.Res.Inst.Math.Sci.Series A, Vol. II (1966), 157-242

[ArWo 68] H.Araki, J.Woods : A Classification of Factors, Publ.Res.Inst. Math.Sci.Series A, Vol. IV (1968), 51-130

[Bau 78] H.Bauer : Wahrscheinlichkeitstheorie und Grundzüge der Masstheorie, 3.Aufl. De Gruyter (1978)

[BaWo 83] H.Baumgärtel, M.Wollenberg : Mathematical Scattering Theory, Akademie Verlag (1983)

[Bla 86] B.Blackadar : K-Theory for Operator Algebras, Springer (1986)

[BöSi 89] A.Böttcher, B.Silbermann : Analysis of Toeplitz Operators, Akademie Verlag (1989)

[Br 77] L.Brown : Stable Isomorphism of Hereditary Subalgebras of C^*-Algebras, Pac. J. Math. 71 (1977), 335-348

[CoHi 68] R.Courant, D.Hilbert : Methoden der Mathematischen Physik I Dritte Aufl. Springer (1968)

[Cu 77] J.Cuntz : Simple C^*-Algebras Generated by Isometries, Com. Math. Phys. 57 (1977), 173-185

[Cu 78] J.Cuntz : Automorphisms of Certain C^*-Algebras, Proc. Conf. Quantum Fields, Algebras, Processes, Bielefeld 1978 (L.Streit ed.)

[Cu 82a] J.Cuntz : K-Theory and C^*-Algebras, Proc. Conf. on K-Theory (Bielefeld Springer) LNM 1046 (1982), 54-79

[Cu 82b] J.Cuntz : The Internal Structure of Simple C^*-Algebras, Proc. Symp. pure Math. 38 AMS (1982), 85-115

[Cu 92] J.Cuntz : A Survey of Some Aspects of Non-Commutative Geometry, preprint (1992)

[Die 84] J.Diestel : Sequences and Series in Banach Spaces, Springer-GTM 92 (1984)

[DiUh 77] J.Diestel, J.Uhl : Vector Measures, AMS-Surveys 15 (1977)

[Dix 77] J.Dixmier : C^*-Algebras, North Holland Amsterdam (1977)

[FaSk 81] T.Fack, G.Skandalis : Connes Analogue of the Thom Isomorphism for the Kasparov Groups, Invent. Math. 64 (1981), 7-14

[Fil 68] P.Filmore : Lectures on Operator Theory, Van Nostrand (1968)

REFERENCES

[Gui 72] A.Guichardet : Symmetric Hilbert Spaces and related Topics, Springer LNM 261 (1972)

[Hal 67] P.Halmos : Lectures on Boolean Algebras, van Nonstrand Math. Studies 1 (1967)

[HeRo II] E.Hewitt, K.Ross : Abstract Harmonic Analysis II, Springer Grundlehren 152 (1970)

[HiPh] E.Hille, S.Phillips : Functional Analysis and Semigroups, AMS Colloquium Publications Vol. 31

[Kir 94] E.Kirchberg : The Classification of Purely Infinite C^*-Algebras using Kasparov's Theory, preprint (1994)

[Lan 73] C.Lance : On Nuclear C^*-Algebras, J. Funct. Anal. 12 (1973), 157-176

[Par 92] K.Parthasarathy : An Introduction to Quantum Stochastic Calculus, Birkhäuser (1992)

[PaSc 72] K.Parthasarathy, K.Schmidt : Positive Definite Kernels, Continuous Tensor Products and Central Limit Theorems of Probability Theory, Springer LNM 272 (1972)

[Ped 79] G.Pedersen : C^*-Algebras and their Automorphism Groups, Academic Press (1979)

[PeTa 79] G.Pedersen, H.Takai : Crossed Products of C^*-Algebras by Approximately Uniformly Continuous Actions, Math.Scand. 45 (1979), 282-288

[Ph 95] C.Phillips : A Classification Theorem for Purely Infinite Nuclear Simple C^*-Algebras, preprint (1995)

[Po 67] R.Powers : Representations of Uniformly Hyperfinite Algebras and their Associated v. Neumann Algebras, Ann. Math. 86 (1967), 138-171

[Po 88] R.Powers : An Index Theory for Semigroups of $*$-Endomorphisms on $\mathcal{B}(H)$ and Type II_1-Factors, Canad. J. Math. 40 (1988), 86-114

[Po 89] R.Powers : A Nonspatial Continuous Semigroup of $*$-Endomorphisms on $\mathcal{B}(H)$, Publ.Res.Inst.Math.Sci. 23 (1989), 1053-1069

[Po 94] R.Powers : New Examples of Continuous Spatial Semigroups on $\mathcal{B}(H)$, preprint (1994)

[PoPr 90] R.Powers, G.Price : Continuous Spatial Semigroups of ∗-Endomorphisms of $\mathcal{B}(H)$, Trans. AMS (1990), 347-361

[PoRo 89] R.Powers, D.Robinson : An Index for Continuous Semigroups of ∗-Endomorphisms of $\mathcal{B}(H)$, J. Funct. Ana. 84 (1989), 85-96

[Rie 82] M.Rieffel : Morita Equivalence for Operator Algebras, Proc. Symp. pure Math. 38 AMS (1982), 285-298

Joachim Zacharias
UFR Faculté des Sciences
Université d'Orléans
Département de Mathématiques
Rue de Chartres - BP 6759
45067 Orléans Cedex 2
France

e-mail : zacharia@labomath.univ-orleans.fr

Editorial Information

To be published in the *Memoirs*, a paper must be correct, new, nontrivial, and significant. Further, it must be well written and of interest to a substantial number of mathematicians. Piecemeal results, such as an inconclusive step toward an unproved major theorem or a minor variation on a known result, are in general not acceptable for publication. *Transactions* Editors shall solicit and encourage publication of worthy papers. Papers appearing in *Memoirs* are generally longer than those appearing in *Transactions* with which it shares an editorial committee.

As of September 30, 1999, the backlog for this journal was approximately 5 volumes. This estimate is the result of dividing the number of manuscripts for this journal in the Providence office that have not yet gone to the printer on the above date by the average number of monographs per volume over the previous twelve months, reduced by the number of issues published in four months (the time necessary for preparing an issue for the printer). (There are 6 volumes per year, each containing at least 4 numbers.)

A Copyright Transfer Agreement is required before a paper will be published in this journal. By submitting a paper to this journal, authors certify that the manuscript has not been submitted to nor is it under consideration for publication by another journal, conference proceedings, or similar publication.

Information for Authors and Editors

Memoirs are printed by photo-offset from camera copy fully prepared by the author. This means that the finished book will look exactly like the copy submitted.

The paper must contain a *descriptive title* and an *abstract* that summarizes the article in language suitable for workers in the general field (algebra, analysis, etc.). The *descriptive title* should be short, but informative; useless or vague phrases such as "some remarks about" or "concerning" should be avoided. The *abstract* should be at least one complete sentence, and at most 300 words. Included with the footnotes to the paper, there should be the 1991 *Mathematics Subject Classification* representing the primary and secondary subjects of the article. This may be followed by a list of *key words and phrases* describing the subject matter of the article and taken from it. A list of the numbers may be found in the annual index of *Mathematical Reviews*, published with the December issue starting in 1990, as well as from the electronic service e-MATH [**telnet e-MATH.ams.org** (or **telnet 130.44.1.100**). Login and password are **e-math**]. For journal abbreviations used in bibliographies, see the list of serials in the latest *Mathematical Reviews* annual index. When the manuscript is submitted, authors should supply the editor with electronic addresses if available. These will be printed after the postal address at the end of each article.

Electronically prepared papers. The AMS encourages submission of electronically prepared papers in $\mathcal{A}_{\mathcal{M}}\mathcal{S}$-TeX or $\mathcal{A}_{\mathcal{M}}\mathcal{S}$-LaTeX. The Society has prepared author packages for each AMS publication. Author packages include instructions for preparing electronic papers, the *AMS Author Handbook*, samples, and a style file that generates the particular design specifications of that publication series for both $\mathcal{A}_{\mathcal{M}}\mathcal{S}$-TeX and $\mathcal{A}_{\mathcal{M}}\mathcal{S}$-LaTeX.

Authors with FTP access may retrieve an author package from the Society's Internet node `e-MATH.ams.org` (130.44.1.100). For those without FTP access, the author package can be obtained free of charge by sending e-mail to `pub@ams.org` (Internet) or from the Publication Division, American Mathematical Society, P.O. Box 6248, Providence, RI 02940-6248. When requesting an author package, please specify $\mathcal{AMS}$-TeX or $\mathcal{AMS}$-LaTeX, Macintosh or IBM (3.5) format, and the publication in which your paper will appear. Please be sure to include your complete mailing address.

Submission of electronic files. At the time of submission, the source file(s) should be sent to the Providence office (this includes any TeX source file, any graphics files, and the DVI or PostScript file).

Before sending the source file, be sure you have proofread your paper carefully. The files you send must be the EXACT files used to generate the proof copy that was accepted for publication. For all publications, authors are required to send a printed copy of their paper, which exactly matches the copy approved for publication, along with any graphics that will appear in the paper.

TeX files may be submitted by email, FTP, or on diskette. The DVI file(s) and PostScript files should be submitted only by FTP or on diskette unless they are encoded properly to submit through e-mail. (DVI files are binary and PostScript files tend to be very large.)

Files sent by electronic mail should be addressed to the Internet address `pub-submit@ams.org`. The subject line of the message should include the publication code to identify it as a Memoir. TeX source files, DVI files, and PostScript files can be transferred over the Internet by FTP to the Internet node `e-math.ams.org` (130.44.1.100).

Electronic graphics. Figures may be submitted to the AMS in an electronic format. The AMS recommends that graphics created electronically be saved in Encapsulated PostScript (EPS) format. This includes graphics originated via a graphics application as well as scanned photographs or other computer-generated images.

If the graphics package used does not support EPS output, the graphics file should be saved in one of the standard graphics formats—such as TIFF, PICT, GIF, etc.—rather than in an application-dependent format. Graphics files submitted in an application-dependent format are not likely to be used. No matter what method was used to produce the graphic, it is necessary to provide a paper copy to the AMS.

Authors using graphics packages for the creation of electronic art should also avoid the use of any lines thinner than 0.5 points in width. Many graphics packages allow the user to specify a "hairline" for a very thin line. Hairlines often look acceptable when proofed on a typical laser printer. However, when produced on a high-resolution laser imagesetter, hairlines become nearly invisible and will be lost entirely in the final printing process.

Screens should be set to values between 15% and 85%. Screens which fall outside of this range are too light or too dark to print correctly.

Any inquiries concerning a paper that has been accepted for publication should be sent directly to the Editorial Department, American Mathematical Society, P. O. Box 6248, Providence, RI 02940-6248.

Editors

This journal is designed particularly for long research papers (and groups of cognate papers) in pure and applied mathematics. Papers intended for publication in the *Memoirs* should be addressed to one of the following editors:

Ordinary differential equations, partial differential equations, and applied mathematics to JOHN MALLET-PARET, Division of Applied Mathematics, Brown University, Providence, RI 02912-9000; electronic mail: `jmp@cfm.brown.edu`.

Harmonic analysis, representation theory, and Lie theory to ROBERT J. STANTON, Department of Mathematics, The Ohio State University, 231 West 18th Avenue, Columbus, OH 43210-1174; electronic mail: `stanton@math.ohio-state.edu`.

Ergodic theory and dynamical systems to ROBERT F. WILLIAMS, Department of Mathematics, University of Texas at Austin, Austin, TX 78712-1082; e-mail: `bob@math.utexas.edu`

Real and harmonic analysis and geometric partial differential equations to WILLIAM BECKNER, Department of Mathematics, University of Texas at Austin, Austin, TX 78712-1082; e-mail: `beckner@math.utexas.edu`.

Algebra to CHARLES CURTIS, Department of Mathematics, University of Oregon, Eugene, OR 97403-1222 e-mail: `cwc@darkwing.uoregon.edu`

Algebraic topology and cohomology of groups to STEWART PRIDDY, Department of Mathematics, Northwestern University, 2033 Sheridan Road, Evanston, IL 60208-2730; e-mail: `s_priddy@math.nwu.edu`.

Differential geometry and global analysis to CHUU-LIAN TERNG, Department of Mathematics, Northeastern University, Huntington Avenue, Boston, MA 02115-5096; e-mail: `terng@neu.edu`.

Probability and statistics to RODRIGO BAÑUELOS, Department of Mathematics, Purdue University, West Lafayette, IN 47907-1968; e-mail: `banuelos@math.purdue.edu`.

Combinatorics and Lie theory to PHILIP J. HANLON, Department of Mathematics, University of Michigan, Ann Arbor, MI 48109-1003; e-mail: `hanlon@math.lsa.umich.edu`.

Logic to THEODORE SLAMAN, Department of Mathematics, University of California at Berkeley, Berkeley, CA 94720-3840; e-mail: `slaman@math.berkeley.edu`.

Number theory and arithmetic algebraic geometry to ALICE SILVERBERG, c/o Mathematisches Institut, Universitaet Erlangen–Nuernberg, Bismarckstraße 1 1/2, 91054 Erlangen, Germany; e-mail: `silver@math.ohio-state.edu`.

Complex analysis and complex geometry to DANIEL M. BURNS, Department of Mathematics, University of Michigan, Ann Arbor, MI 48109-1003; e-mail: `dburns@math.lsa.umich.edu`.

Algebraic geometry and commutative algebra to LAWRENCE EIN, Department of Mathematics, University of Illinois, 851 S. Morgan (M/C 249), Chicago, IL 60607-7045; e-mail: `ein@uic.edu`.

Geometric topology, knot theory, hyperbolic geometry, and general topoogy to JOHN LUECKE, Department of Mathematics, University of Texas at Austin, Austin, TX 78712-1082; e-mail: `luecke@math.utexas.edu`.

Partial differential equations and applied mathematics to BARBARA LEE KEYFITZ, Department of Mathematics, University of Houston, 4800 Calhoun, Houston, TX 77204-3476; e-mail: `keyfitz@uh.edu`

Operator algebras and functional analysis to BRUCE E. BLACKADAR, Department of Mathematics, University of Nevada, Reno, NV 89557; e-mail: `bruceb@math.unr.edu`

All other communications to the editors should be addressed to the Managing Editor, PETER SHALEN, Department of Mathematics, University of Illinois, 851 S. Morgan (M/C 249), Chicago, IL 60607-7045; e-mail: `shalen@math.uic.edu`.

Selected Titles in This Series

(Continued from the front of this publication)

649 **Bernd Stellmacher and Franz Georg Timmesfeld,** Rank 3 amalgams, 1998

648 **Raúl E. Curto and Lawrence A. Fialkow,** Flat extensions of positive moment matrices: Recursively generated relations, 1998

647 **Wenxian Shen and Yingfei Yi,** Almost automorphic and almost periodic dynamics in skew-product semiflows, 1998

646 **Russell Johnson and Mahesh Nerurkar,** Controllability, stabilization, and the regulator problem for random differential systems, 1998

645 **Peter W. Bates, Kening Lu, and Chongchun Zeng,** Existence and persistence of invariant manifolds for semiflows in Banach space, 1998

644 **Michael David Weiner,** Bosonic construction of vertex operator para-algebras from symplectic affine Kac-Moody algebras, 1998

643 **Józef Dodziuk and Jay Jorgenson,** Spectral asymptotics on degenerating hyperbolic 3-manifolds, 1998

642 **Chu Wenchang,** Basic almost-poised hypergeometric series, 1998

641 **W. Bulla, F. Gesztesy, H. Holden, and G. Teschl,** Algebro-geometric quasi-periodic finite-gap solutions of the Toda and Kac-van Moerbeke hierarchies, 1998

640 **Xingde Dai and David R. Larson,** Wandering vectors for unitary systems and orthogonal wavelets, 1998

639 **Joan C. Artés, Robert E. Kooij, and Jaume Llibre,** Structurally stable quadratic vector fields, 1998

638 **Gunnar Fløystad,** Higher initial ideals of homogeneous ideals, 1998

637 **Thomáš Gedeon,** Cyclic feedback systems, 1998

636 **Ching-Chau Yu,** Nonlinear eigenvalues and analytic-hypoellipticity, 1998

635 **Magdy Assem,** On stability and endoscopic transfer of unipotent orbital integrals on p-adic symplectic groups, 1998

634 **Darrin D. Frey,** Conjugacy of Alt_5 and $SL(2,5)$ subgroups of $E_8(\mathbb{C})$, 1998

633 **Dikran Dikranjan and Dmitri Shakhmatov,** Algebraic structure of pseudocompact groups, 1998

632 **Shouchuan Hu and Nikolaos S. Papageorgiou,** Time-dependent subdifferential evolution inclusions and optimal control, 1998

631 **Ronnie Lee, Steven H. Weintraub, and J. William Hoffman,** The Siegel modular variety of degree two and level four/Cohomology of the Siegel modular group of degree two and level four, 1998

630 **Florin Rădulescu,** The Γ-equivariant form of the Berezin quantization of the upper half plane, 1998

629 **Richard B. Sowers,** Short-time geometry of random heat kernels, 1998

628 **Christopher K. McCord, Kenneth R. Meyer, and Quidong Wang,** The integral manifolds of the three body problem, 1998

627 **Roland Speicher,** Combinatorial theory of the free product with amalgamation and operator-valued free probability theory, 1998

626 **Mikhail Borovoi,** Abelian Galois cohomology of reductive groups, 1998

625 **George Xian-Zhi Yuan,** The study of minimax inequalities and applications to economies and variational inequalities, 1998

624 **P. Deift and K. T-R McLaughlin,** A continuum limit of the Toda lattice, 1998

623 **S. A. Adeleke and Peter M. Neumann,** Relations related to betweenness: Their structure and automorphisms, 1998

For a complete list of titles in this series, visit the
AMS Bookstore at **www.ams.org/bookstore/**.